EXPERIMENTAL ELECTRONICS FOR STUDENTS

EXPERIMENTAL ELECTRONICS FOR STUDENTS

K. J. Close
Principal Lecturer,
School of Engineering and Science,
The Polytechnic of Central London

and

J. Yarwood
Formerly Head of Physics,
School of Engineering and Science,
The Polytechnic of Central London

Errata
In two of the circuit diagrams presented in this book errors have occurred in the figures.

Figure 3.1 (page 27): the battery symbol on the left is shown incorrectly. The polarity should be reversed.

Figure 5.10 (page 101): the diode D_2 is shown incorrectly. The symbol should be reversed.

Experimental Electronics for Students: Close and Yarwood

LONDON
CHAPMAN AND HALL
A Halsted Press Book
John Wiley & Sons, New York

First published 1979
by Chapman and Hall Ltd.
11 New Fetter Lane, London EC4P 4EE

Printed in Great Britain by
M. & A. Thomson Litho Ltd., East Kilbride, Scotland

Distributed in the U.S.A. by Halsted Press,
a Division of John Wiley & Sons, Inc., New York

British Library Cataloguing in Publication Data

Close, Kenneth John
Experimental electronics for students.
1. Electronics — Laboratory manuals
I. Title II. Yarwood, John
621.381:028 TK7818

ISBN 0-412-14760-2

U.S.A. Halsted Press ISBN 0-470-26767-4 cased

Preface

Electronics is essentially an experimental subject and enables a wealth of experimental work to be undertaken at relatively low cost. In any modestly equipped electrical engineering or physics laboratory, it is possible to plan interesting experiments to study active and passive components, basic circuit functions, modular encapsulations and monolithic integrated circuits. The work may range from the formal investigation of a device new to the student to the design and construction of quite advanced, modern measurement and control systems.

There are few books which guide experimental work in electronics. This text aims to rectify this by giving detailed descriptions of a series of experiments all of which have been thoroughly tested by students in physics, electronics, electrical engineering and instrumentation at The Polytechnic of Central London. Moreover, several of these experiments would seem to be appropriate for the current development of interest in courses in electronics in schools because several of them have been undertaken with considerable success by first-year sixth-form students who have come to Central London for special courses. They would also assist an introductory course in electronics for students from other disciplines and have been tried out in this way at The Polytechnic.

The authors are convinced that experimental work in

electronics should be introduced at an early stage in schools, certainly in sixth-form work if not at more elementary stages. The reasons for this are fourfold: the subject forms a natural extension to elementary electric circuit practice; it has an inherent fascination for many students; the devices and components are readily available at low cost and it enables the student to gain a real insight into present-day practice in science and technology.

In many of the experiments described sufficient background information has been provided for a student to proceed without a formal lecture. This makes a self-study programme easy to plan. Teachers will find this experimental guide helpful when introducing a new circuit configuration or planning project work.

It is hoped that the ideas and suggestions contained in this book will be of value in the many courses involving electronics in secondary and tertiary education.

London, 1979.

K J C
J Y

Contents

CHAPTER ONE

Introductory general notes

1.1 *Soldering components to interconnections when utilizing strip-board*

Most of the experiments described in Chapter two onwards are based on circuits. To construct many of these circuits it is often good practice to make us of Veroboard (a trade name for a type of strip-board) on which the passive and active components may be readily mounted and their leads soldered to the appropriate interconnections already provided on the strip-board.

Veroboard (Fig. 1.1a) is a plastic board to one side of which are bonded parallel copper strips (this side is called the 'copper side'). Tiny holes are spaced at equal intervals along the copper strips. The spacing between these holes is 0.1 inch in boards to which are attached integrated circuit (IC) modules (0.1 inch is the separation between the pins of integrated circuits) whereas this spacing is 0.15 inch for general use, and is usually preferable. The holes are of small diameter, just big enough to allow leads through to the components. As shown in Fig. 1.1b, these components (passive, active, integrated circuit modules) are placed on one side of the board with their leads bent over to pass through appropriate holes to the other side of the board (the copper tracks). All leads soldered to a given track are obviously interconnected electrically; however, a track can

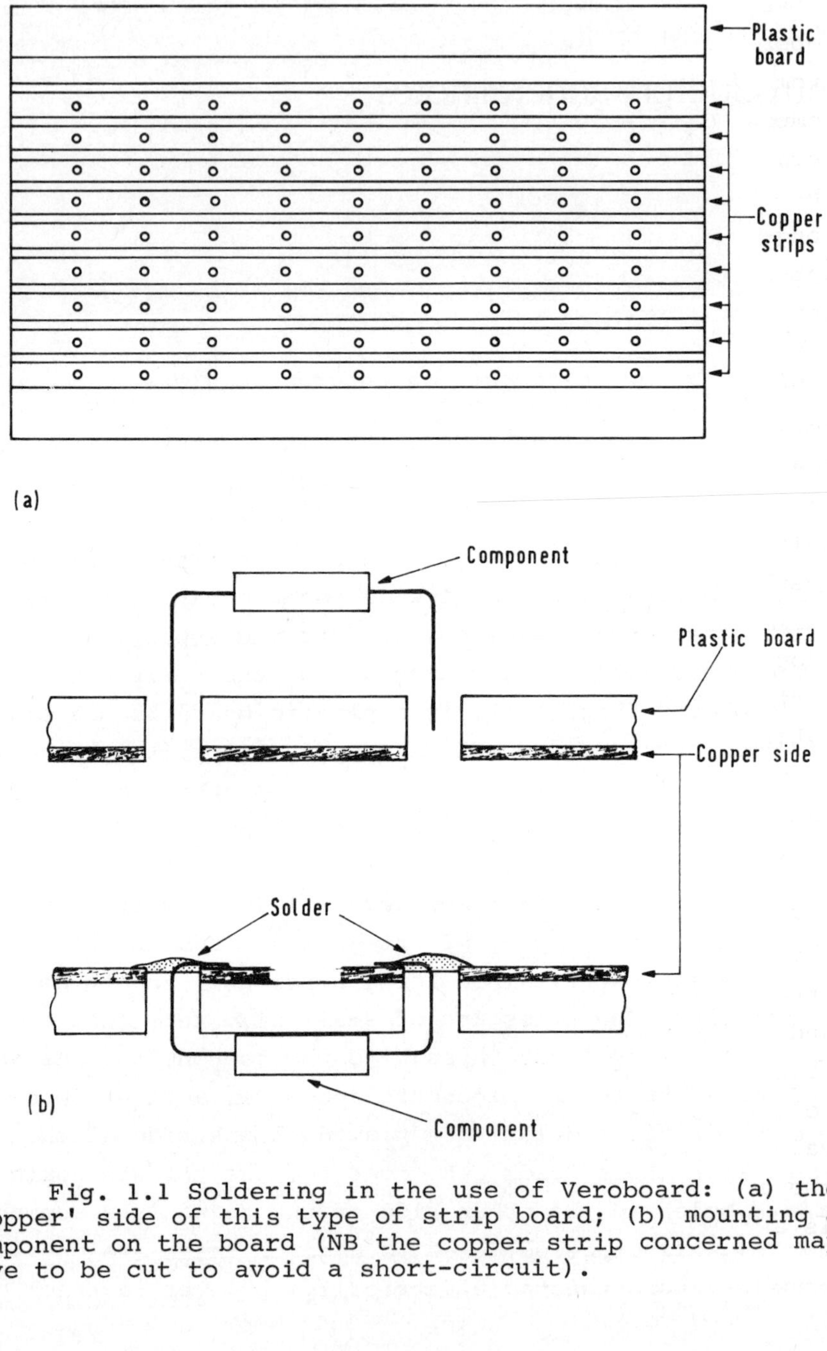

Fig. 1.1 Soldering in the use of Veroboard: (a) the 'copper' side of this type of strip board; (b) mounting a component on the board (NB the copper strip concerned may have to be cut to avoid a short-circuit).

be cut into two or more isolated sections if needed.

As the components are almost always miniature, it is important to minimize the amount of heat that is conveyed from the soldering-iron tip (to be kept clean) to the component; otherwise the small component may be ruined because it becomes heated to an excessive temperature. The lead being soldered is therefore best gripped with long-nose pliers (as close to the device as possible) to act as a sink to conduct heat away from the component. Once the component has been placed in position on the strip-board the soldering-iron is applied for a few seconds and then a small quantity of solder (best rosin-cored solder) is applied on the copper side of the strip-board to connect the component leads to the copper track; the solder should never be applied and carried on the soldering-iron tip itself. The solder should 'run' and 'wet' the surfaces to be joined.

It is clearly essential to decide the layout to be adopted in mounting components on Veroboard so that a particular circuit is constructed in the most satisfactory way. It is preferable to plan this layout on paper first (unless very experienced) and then undertake the actual layout and soldering of component leads on the Veroboard.

1.2 *Resistor colour code*

Dimensionally small resistors are of two main constructional types: carbon film (providing high stability); and metal oxide. For the former the E-band (Fig. 1.2) is coloured pink; for the metal oxide type, the E-band is coloured black. To enable ready identification of the value of the resistor in ohms, the code used to determine the colour of the other bands, A, B, C and D (Fig. 1.2) is as given in Table 1.1.

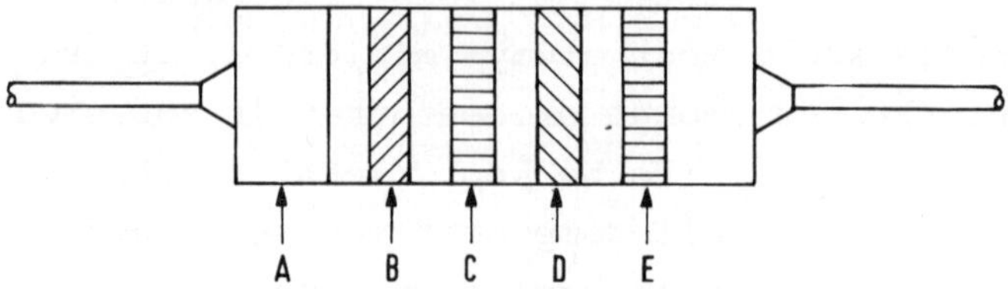

Fig. 1.2 Resistor code

Table 1.1 Resistor colour code

Colour	A 1st digit of resistance (in ohms)	B 2nd digit of resistance (in ohms)	C Multiplier of resis- tance value	D Tolerance (%)
Black	0	0	1	-
Brown	1	1	10^1	1%
Red	2	2	10^2	2%
Orange	3	3	10^3	-
Yellow	4	4	10^4	-
Green	5	5	10^5	-
Blue	6	6	10^6	-
Violet	7	7	10^7	-
Grey	8	8	-	-
White	9	9	-	-
Gold	-	-	10^{-1}	5%
Silver	-	-	10^{-2}	10%
None	-	-	-	20%

Note: order of colours Red to Violet is as in the spectrum.

Example: for a certain resistor colours of rings are A-orange; B-brown, C-yellow, D-red. Reference to Table 1.1 shows that its resistance is 31×10^4 ohm = 310 kΩ and the tolerance of this value is 2%.

1.3 *Symbols used in circuit diagrams*

The symbols used in the text for passive components, active components and connectors follow the recommended practice of the British Standards Institution (BSI) and the International Electrotechnical Commission (IEC). These symbols for components are the subject of Fig. 1.3, whereas for the differential amplifier and logic functions see Fig 1.4.

Within the text the usual decimal notation is adopted in specifying the values of passive components, eg 5.6kΩ, 2.5 μF, 32.0 mH whereas on circuit diagrams the practice recommended by the standards institutions, which has been followed, is to avoid the use of decimal points. The resistance values given in Table 1.2 clarify this procedure. In the case of capacitance, for example, if 50 μF appears in the text, this will be simply 50 on the circuit diagram. On the other hand, if the value of the capacitance is better given in some sub-division of the farad other than the microfarad, for example, the picofarad, then the diagram and the text will both have, say, 50 pF.

Table 1.2 Specification in the text and on diagrams of resistance values

Resistance value in text (Ω)	Often written in text	Sometimes written in text	Recommended by BSI for use on diagrams
2 200 000	2.2 MΩ	2.2M	2M2
4 700	4.7 kΩ	4.7k	4k7
33	33 Ω	33	33
4.7	4.7 Ω	4.7	4R7

1.4 *Symbols for quantities*

Conventional standards practice is followed here. In general, lower case italic letters are used for varying quantities and capital italic letters for steady quantities, eg a voltage which varies with time is denoted by v, whereas a voltage which does not vary with time during the experiment is denoted by V.

1.5 *Abbreviations*

Modern practice is followed of omitting full-stops after

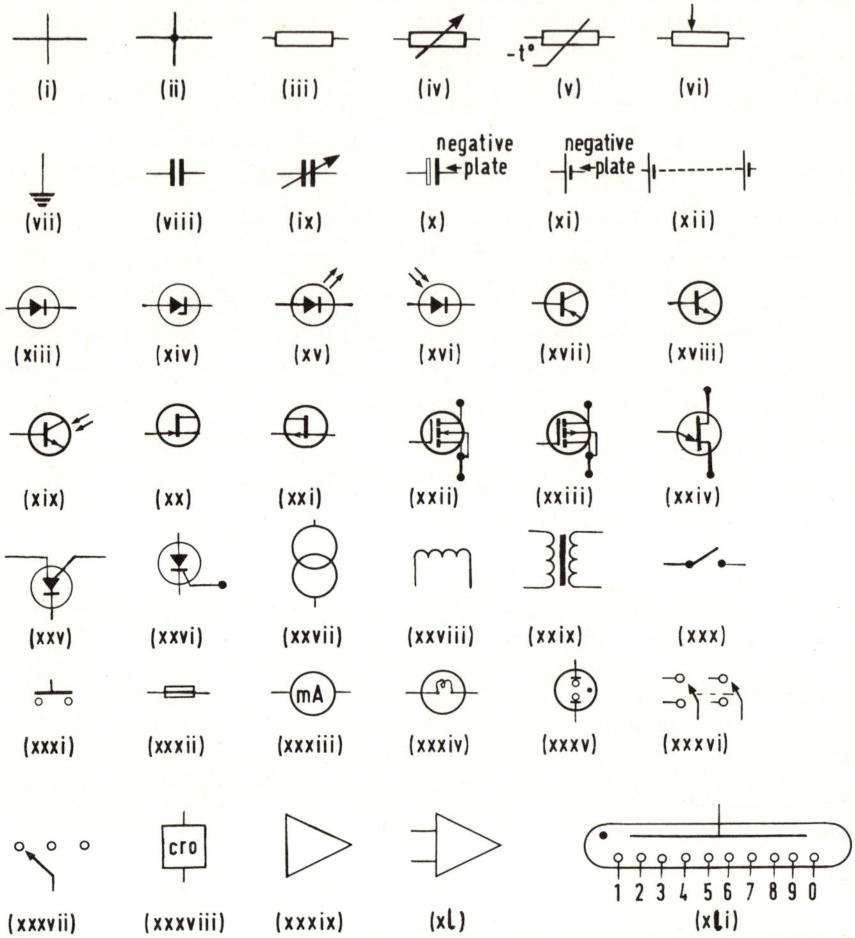

Fig. 1.3 Component symbols in circuit diagrams: (i) crossing connectors; (ii) joined connectors (NB two separated dots are usual on the horizontal line with two separated vertical lines instead of one with each terminating at its own dot on the horizontal line); (iii) resistor (fixed); (iv) resistor (variable); (v) thermistor; (vi) potentiometer; (vii) earth (or ground); (viii) capacitor (fixed); (ix) capacitor (variable); (x) electrolytic capacitor; (xi) single-cell battery; (xii) multi-cell battery; (xiii) diode; (xiv) Zener diode; (xv) light-emitting diode; (xvi) photo-diode; (xvii) bipolar transistor (p-n-p); (xviii) bipolar transistor (n-p-n); (xix) photo-transistor; (xx) n-channel jgfet; (xxi) p-channel jgfet; (xxii) n-channel enhancement igfet; (xxiii) p-channel enhancement igfet; (xxiv) unijunction transistor; (xxv) programmable unijunction transistor;

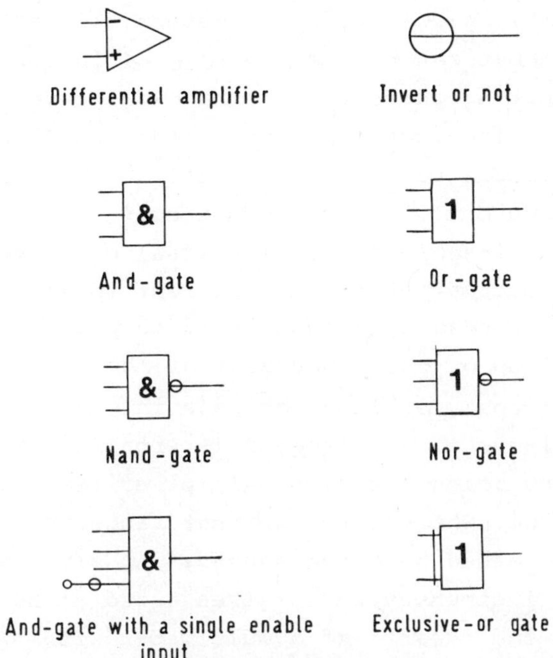

Fig. 1.4 Differential amplifier and logic function symbols.

the accepted abbreviation, thus 'for example' is not written e.g. but eg and alternating current is not written a.c. but ac. Furthermore, except where inappropriate, lower case letters are generally used rather than capital letters.

1.6 *Notes on some aspects of electrical measuring istruments*

Moving-coil instruments are comparatively cheap, widely-used and lend themselves to the development of multi-

---

(xxvi) silicon-controlled rectifier; (xxvii) current generator; (xxviii) inductor; (xxix) iron-cored transformer; (xxx) on-off switch; (xxi) singlepole switch, push to make; (xxxii) fuse; (xxxiii) lettering shows type of instrument (xxxiv) filament lamp; (xxxv) indicator lamp; (xxxvi) double-pole, two-way switch (xxxvii) single-pole three-way switch; (xxxviii) cathode ray oscillograph; (xxxix) amplifier; (xl) differential amplifier; (xli) number tube.

meters. The moving-coil meter is essentially actuated by direct current but can be readily adapted for the measurement of alternating current (of sinusoidal waveform) by the appropriate incorporation of a diode rectifier, usually in a bridge rectifier circuit.

A direct current of 50 microamperes (50 μA) is needed to produce a full-scale deflection (fsd) of a sensitive panel-mounted microammeter. Any current smaller than 50 μA cannot be readily accommodated to provide a fsd without sacrifice of the robustness of the meter, considerably greater cost, or inconvenience in use.

As is well-known, a voltmeter is obtained from a microammeter by using a series multiplier resistance of calculated value, whereas the current range of the microammeter is increased by using shunts. As a voltmeter, a moving-coil instrument which gives a fsd at 50 μA (which could well be the 'heart' of a multimeter with switchable voltage and current ranges, including alternating ones) will have a resistance of 20 000 ohms per volt (20 kΩ V^{-1}). This represents a high resistance voltmeter in moving-coil practice.

However, in several instances of measurements on electronic circuits, a high resistance voltmeter of this kind may demand too great a current from the circuit, even though this current is only 50 μA at fsd. If this is the case, the considerably more expensive digital voltmeter (dvm or DVM) is recommended. A good quality dvm can have a resistance as high as 10 megohm (10 MΩ) or even 100 MΩ and so draw negligible current from the circuit; it is able to give an accuracy of reading of about 0.01%, compared with about 0.1% for a moving-coil instrument. The dvm can also be used to measure current by virtue of its measurement of the pd produced across a known resistance through which this current is passed.

If available, it is recommended that a dvm be used to

measure pd in the experiments; digital multimeters are also made and are very useful. Failing availability of a suitable dvm, any moving-coil voltmeter used should have a resistance of 20 000 Ω V^{-1} or more. If a voltmeter of lower resistance than this is used, care must be taken as to how it is connected across a circuit otherwise it may introduce serious errors in measurement, because the current it demands is a significant fraction of the current through any parallel high resistance circuit across it.

The use of a cathode-ray oscilloscope (cro or CRO) is referred to repeatedly in the experimental work. A cro is able, of course to delineate on its screen the waveform of a voltage or current. This screen trace can give a permanent record either by tracing or photography. If the cheaper tracing method is adopted, all that is needed is to place a piece of transparent graph paper against the fluorescent screen of the cro and trace the waveform on it with a suitable pencil or fibre pen. Photography is more expensive and can be undertaken in one of the following two ways:

(i) Use a camera (eg a 35 mm film camera or, preferably the camera supplied by the manufacturer of the cro) to make a photographic negative of the fluorescent screen trace, develop this negative and then make from it a suitable print or enlargement.

(ii) Use a Polaroid camera again preferably as supplied by the manufacturer for the cro in question: this is quicker but more expensive.

In several of the experiments it is stated that the waveform is 'displayed and recorded by a cro'. Usually this cro is of the double-beam variety so that two waveforms can be shown simultaneously - valuable when it is needed to compare then, eg from the point of view of the phase difference between them. Moreover, the cro is provided with amplifiers to both y-inputs and, sometimes, has an x-amplifier. So 'displayed and recorded' means that the waveform on the fluorescent screen is viewed by eye, brought to a suitable state by adjustment, as necessary, of the focusing, brilliance, time-base, synchronization and cro amplifier gain controls and, then, if a permanent record is required, this waveform is either traced or photographed, where the latter is preferred if the laboratory has the necessary equipment and funds.

It is frequently useful to have a cro (double beam for preference) which has also the following facilities:

(a) A calibrated time-base, ie such that at each setting the time of the total sweep of the beam from left-to-right is known and it is also known that the horizontal sweep is linear. This enables the time period of pulses and of oscillations to be determined either directly from the screen trace or from the record obtained.

(b) y-deflection voltage sensitivity and a y-amplifier of known voltage gain. This enables the vertical deflection in a trace to be converted directly into voltage and recorded as such. With a double-beam cro it is valuable to have this feature with both y_1 and y_2.

If the cro available is not provided with the facilities (a) and (b) it is usually not difficult to provide one's own calibration by making use in case (a) of a variable oscillator (signal generator) with a voltage output of known frequency and in case (b) of known (or measured) sources of voltage.

It should be stressed that to measure accurately the pulse repetition rate (ppr) of a generator, it is generally better to make use of a digital frequency meter rather than a calibrated cro time-base.

When using a cro it should be borne in mind that the frequency response may sometimes be inadequate to record satisfactorily the rise or fall of a pulse which is very rapid, eg of a pulse of rectangular shape. Deficiency in the frequency response of a cro is usually due to inadequacy of its *y*-amplifier. The remedy, generally, is to obtain a more expensive cro.

CHAPTER TWO

Semiconductor diodes: characteristics: use in D.C. power supplies

2.1 *Semiconductor diodes*

The current I which flows through a p-n junction is related to the pd V which exists across the junction by the equation

$$I = I_s\ (\exp(eV/kT) - 1) \qquad (2.1)$$

where I_s is the reverse saturation current, e is the electronic charge, k is the Boltzmann constant and T is the junction temperature on the absolute scale. At T = 300 K, using SI units,

$$e/kT = 1.6 \times 10^{-19}/(1.38 \times 10^{-23} \times 300) \simeq 38\ V^{-1}$$

When the junction is forward biased, V is positive. If V is positive and exceeds a few hundred millivolts.in magnitude, equation (2.1) becomes, approximately,

$$I = I_s \exp\ (eV/kT) \qquad (2.2)$$

When the junction is reverse biased, V is negative; when V exceeds a few hundred millivolts in magnitude, equation (2.1) becomes approximately

$$I = I_s \qquad (2.3)$$

where I_s is constant.

The typical characteristics of germanium and silicon p-n junction diodes are hence as shown in Fig. 2.1.

To enable the electrodes to be identified, the manufacturer often prints the diode circuit symbol on the body of the diode. Sometimes, a ring is marked around the body close to the cathode, ie the electrode to be made negative when the diode is forward biased.

2.1.1 *Characteristics of p-n junction diode.* Fig. 2.2 shows the circuits for determining the characteristics of a semiconductor p-n junction diode in which (a) is for the forward characteristic and (b) is for the reverse characteristic.

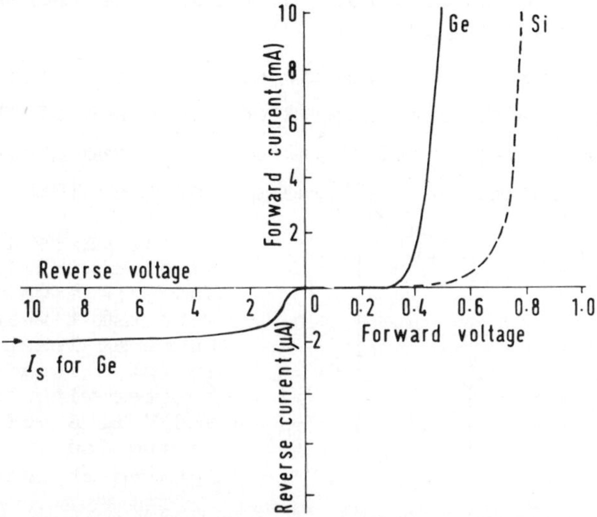

Fig. 2.1 The typical characteristic of p-n junction diode: Ge: germanium; Si: silicon.

Note that in (a) the forward current is measured by a milliammeter whereas in (b) the reverse current is measured by a microammeter.

The circuits are used to determine the characteristics of (i) a germanium point contact diode and (ii) a silicon diode. In case (i) with the germanium diode forward-biased (Fig. 2.2a) the pd across the diode is increased from zero in steps of 0.1 V and at each setting the current through the diode is recorded by the milliammeter where this forward current should not be allowed to exceed 10 mA. For the portion of the forward characteristic close to the origin (ie for small values of the pd V) a micro-ammeter will be necessary in place of the milliammeter:

the convenient choice here is a suitable multimeter. With the germanium diode reverse-biased (Fig. 2.2b), the reverse current is measured by the microammeter for pds across the p-n junction of 0.9 V.

Having obtained the results required for the reverse characteristic, the germanium diode may be warmed with the finger or by means of a hair drier and the increase in reverse current noted.

The procedure involving the circuits of Figs. 2.2a and b is repeated with a silicon diode in place of the germanium one. The characteristics for those two diodes are preferably plotted on the same set of axes (Fig. 2.3).

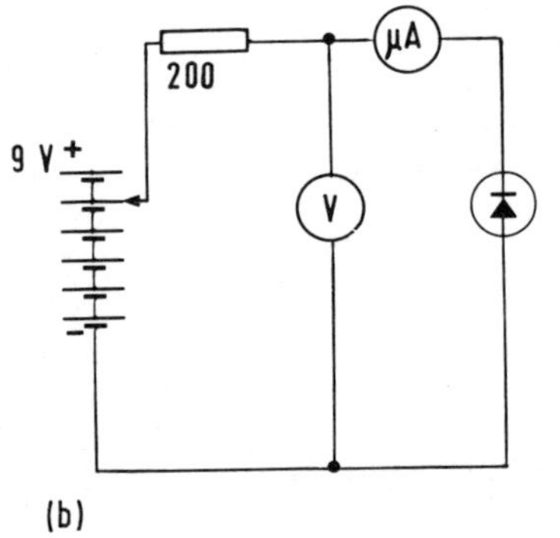

In (a) mA is a milliammeter with fsd 10 mA and V is a voltmeter with fsd 1 V and resistance at least 10kΩV^{-1} in (b) A is a microammeter with fsd 100 μA and V is a voltmeter having fsd 10 V and resistance at least 10kΩV^{-1}

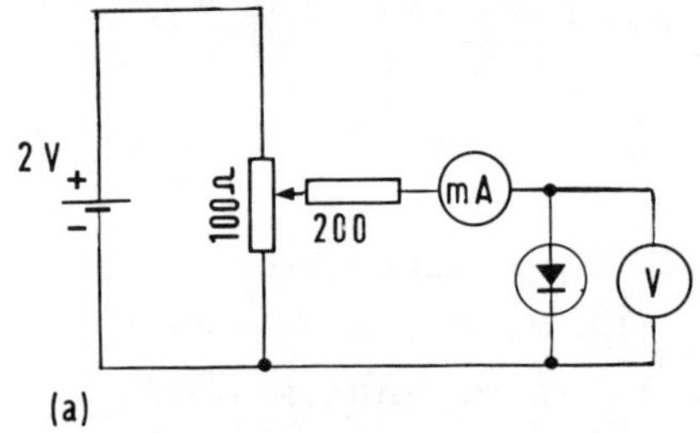

Fig. 2.2 Determination of the characteristics of a p-n junction diode with (a) forward bias and (b) reverse bias.

2.1.2 *Determination of e/k.* Equation (2.2) is conveniently written in logarithmic form

$$ln\ I = ln\ I_s + eV/kT$$

For the germanium point contact diode, plot a graph on log-linear paper of *ln I* against *V*. From the slope of

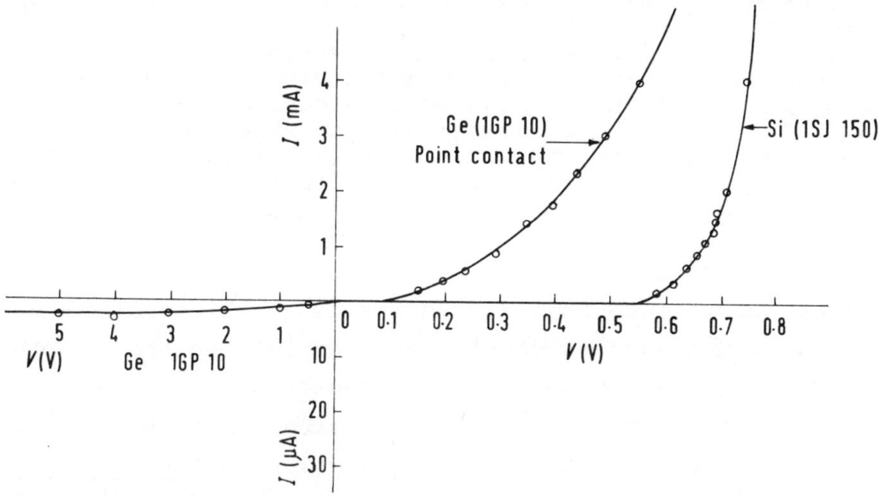

Fig. 2.3 Typical characteristics of a germanium point contact diode and of a silicon diode

Note: the reverse saturation current (ie the leakage current) through a silicon diode is much too small to be measured with a moving-coil microammeter (0 - 100 μA). Indeed, if a current through the silicon diode is detected when it is reverse biased, the diode must be checked to find out if it is faulty or if the current measured is flowing through the voltmeter because it is wrongly connected.

this graph, for a known value of the temperature T, e/k (the ratio of two fundamental constants) can be found.

2.1.3 *Questions*

(a) What features of the characteristic enable an immediate distinction to be made between the germanium and silicon devices?

(b) Which diode would be more efficient as a rectifier?

(c) Why does the reverse saturation current in the germanium diode increase as the junction temperature is raised?

2.1.4 *Further investigations*

(a) Germanium junction diodes are now fairly rare. A simple way of obtaining the equivalent is to join together electrically the collector and base electrodes of a germanium bipolar junction transistor and plot the forward and reverse characteristics of the resulting diode.

A transistor used in this manner is often used to provide a logarithmic characteristic.

(b) Plot the forward and the reverse characteristics of a germanium diode at 30 °C and at 40 °C; such temperatures are conveniently obtained in a constant temperature bath of a heated maize oil or sunflower seed oil and are measured by a simple mercury-in-glass thermometer.

(c) Describe the use of a diode to protect a delicate meter-movement.

2.2 *Zener diodes*

A Zener or voltage regulator diode is a silicon diode specially designed to operate continuously in the reverse-biased mode beyond the breakdown region. If the p-n junction is to survive and the behaviour is to be reproducible, the current through the junction must be limited by a means of a series resistance. A Zener diode is NEVER operated without a suitable resistance connected in series; the magnitude of this resistance is determined knowing the maximum power which may be dissipated safely in the junction.

The Zener voltage V_Z usually marked on the body of the Zener diode, is specified by the manufacturer within ± 10%, eg a diode marked 15 V could have a value of V_Z between 13.5 V and 16.5 V.

When planning to use a Zener diode certain information is essential.

(i) The Zener or turn-over voltage V_Z (Fig. 2.4).

(ii) The minimum current $I_{Z\ min}$ and the maximum current current $I_{Z\ max}$ which can be passed through the diode. The minimum current I_{Zmin} flows when the current through the load connected across the diode is a maximum; $I_{Z\ min}$ has to be chosen so that the operation of the Zener diode is clear of the knee in the characteristic (Fig. 2.4); in an ideal Zener diode, this knee is sharp. The maximum current $I_{Z\ max}$ flows when the load current is zero; it is caluclated knowing the maximum power that can be dissipated safely in the junction.

(iii) The temperature coefficient of V_Z, ie dV_Z/dT. For diodes with $V_Z < 5V$, the Zener voltage decreases as the temperature increases, so dV_Z/dT is negative. For diodes with $V_Z > 6$ V, dV_Z/dT is positive. Consequently, there exists a region between 5 V and 6 V in which dV_Z/dT is zero or near zero, ie V_Z is almost independent of temperature. When high stability against temperature change is required two 5.6 V Zener diodes may be connected in series to provide 11.2 V with a much lower dV_Z/dT than a single 11.2 V Zener diode.

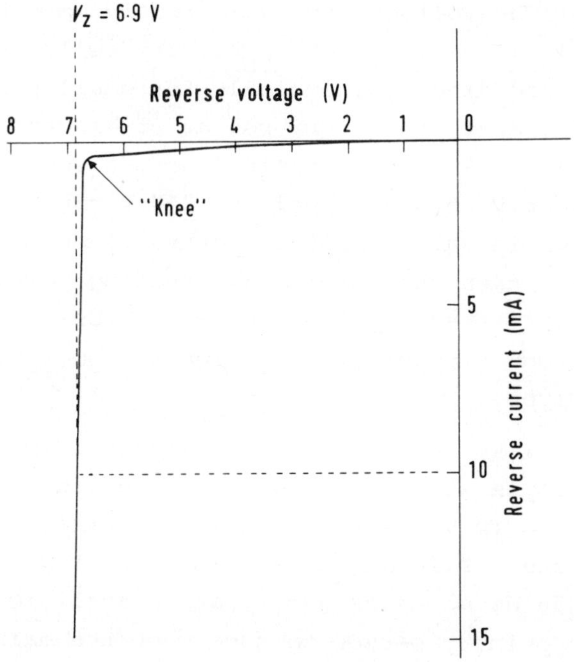

Fig. 2.4 The reverse characteristic of a Zener diode.

2.2.1 *The reverse characteristic of a Zener diode.* The circuit shown in Fig. 2.5 is used. The current through the Zener is not allowed to exceed 15 mA; the slope resistance dV_Z/dI_Z is determined at I_Z = 10 mA.

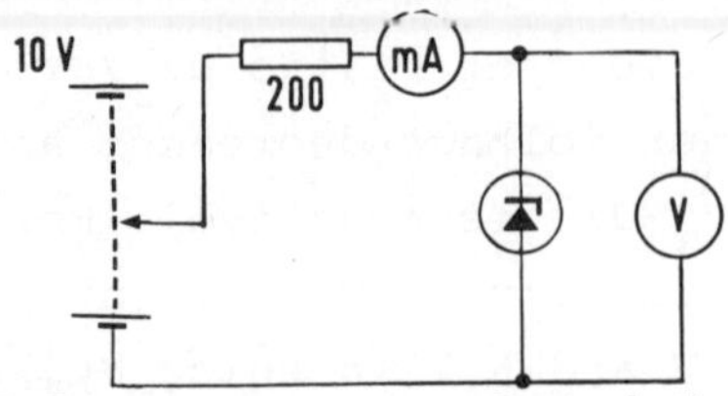

Fig. 2.5 To determine the characteristic of a Zener diode. mA is a milliammeter with fsd 20 mA and V is a voltmeter with fsd of 10 V; a dvm is preferred

2.2.2 *Question.* What 'slope resistance' would you expect a perfect voltage regulating diode to have?

2.3 *D.C. power supplies: an introduction*

Electric power with a steady output voltage (ie a dc power supply) is required to operate almost all electronic equip-equipment. The voltage needed is rarely more than 30 V and commonly is 5 V, 9 V or 12 V. Dry cells (small batteries) are used, particularly for small portable equipment but the most convenient source of electric power is the 240 V rms, 50 Hz alternating current supply from the mains. Clearly this ac supply at 240 V rms must be stepped-down in value, most conveniently by a step-down transformer, then the smaller alternating voltage converted into a steady voltage supply, which requires rectification and smoothing. The simplest rectifier is the half-wave type.

2.3.1 *Half-wave rectifier.* The circuit (Fig. 2.6a) comprises a simple step-down transformer having a 240 V rms primary and a 12 V rms secondary; the primary, with a switch S_1 and a fuse in series, is connected across the 240 V rms 50 Hz ac mains supply and a small neon lamp across the primary serves to show when the mains supply is on; the secondary is connected across a resistor R (from a resistance box) with a semiconductor diode in series and a cro across R. The semiconductor diode can be short-circuited by closing the switch S_2. One end of the secondary winding is earthed, a convenient and necessary feature as a cro is used to observe voltage waveforms.

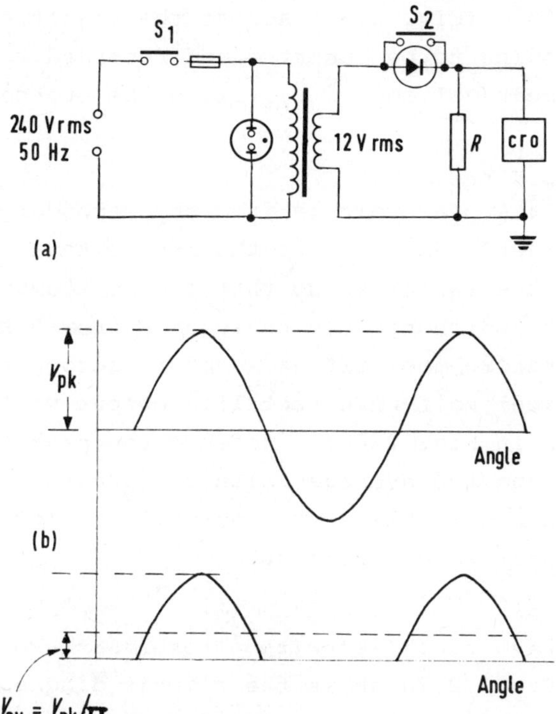

Fig. 2.6 (a) Secondary of a transformer across rectifier (with shorting switch S_2), resistor R and cro. (b) Sinusoidal waveform observed if S_2 is closed; half-wave rectified waveform observed if S_2 is open.

A suggested procedure (where the stages (a) and (b) can be omitted by those familiar with mains supplies) to

(a) Examine the mains plug; at present it is almost invariably of the 13 A fused type and where the correct colour coding is

Yellow-green	-	earth (E)
Brown	-	live (L)
Blue	-	neutral (N)

(b) Set-up the circuit of Fig. 2.6a with the switch S_2 closed so that it consists simply of the secondary winding (12 V) across a resistor R of 5 kΩ (eg from a resistance box) and cro. This enables the sinusoidal voltage waveform (Fig. 2.6b) across the transformer secondary winding to be observed and recorded. Measure the peak-to-peak voltage (2 V_{pk}) with the pre-calibrated cro. Then

$$\sqrt{2}\ V_{rms} = V_{pk}$$

where V_{pk} is the peak voltage from the secondary winding alternating supply and V_{rms} is the rms voltage.

(c) Open the switch S_2 so that the semiconductor diode is in circuit and change the resistor R from 5 kΩ to 2 kΩ. Observe and record the voltage waveform across the resistor (the typical half-wave rectifier output voltage waveform is shown in Fig. 2.6b). Measure the peak voltage V_{pk} with the cro and the average voltage V_{av} across the 2 kΩ resistor with a suitable dc moving-coil voltmeter. Check that, for a half-wave rectifier

$$V_{av} = V_{pk}/\pi$$

2.3.2 *Half-wave rectifier with a reservoir smoothing capacitor*. Fig. 2.7a shows the circuit diagram of a half-wave rectifier with the addition of a capacitor C (an electrolytic capacitor of 500 μF is suitable) across the load resistor R. It is useful if this load resistor R is variable (conveniently supplied by a resistance box). With the cro observe and record the output voltage waveform across R and measure the peak-to-peak ripple voltage (Fig. 2.7b) for various values of R between 3 kΩ and 10 kΩ To illustrate the ripple voltage, Fig. 2.7b shows a simplified approach in which it is assumed that the capacitor charges up to the peak voltage instantaneously and discharges at a uniform rate determined by the load current. The average current through R can be determined by measuring the voltage across it with a suitable moving-coil voltmeter.

Calculate the voltage drop which would occur if the above assumptions were valid and compare the values with the measured peak-to-peak ripple voltage.

Example $C = 500\ \mu F$ $R = 5\ k\Omega$

Average voltage across R = 58 V; average current = 58/5 = 11.6 mA.

Discharging uniformly for 0.2 s, the charge lost

$$= 11.6 \times 10^{-3} \times 2 \times 10^{-2}\ C$$

The drop in voltage

$$Q/C = 23.2 \times 10^{-5}/(500 \times 10^{-6}) = 0.46\ V$$

The measured peak-to-peak ripple voltage = 0.45 V).

Note that this use of a half-wave rectifier with a smoothing capacitor has effectively converted the ac mains supply voltage into a steady voltage of smaller magnitude. Further, in the example just given, the peak-to-peak ripple voltage of 0.45 V is only (0.45/58) × 100 = 0.76% of the average voltage across the load resistor.

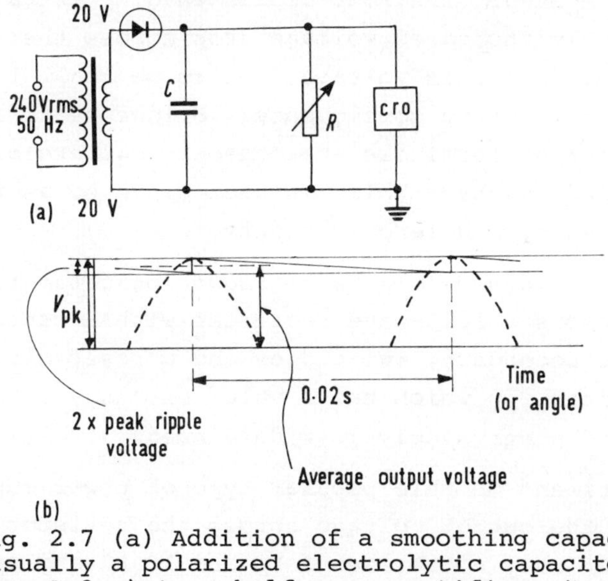

Fig. 2.7 (a) Addition of a smoothing capacitor C (usually a polarized electrolytic capacitor, see Fig. 1.3 x) to a half-wave rectifier circuit; (b) simplified diagram illustrating effect on a half-wave rectifier voltage waveform of a smoothing capacitor.

2.3.3 *Full-wave rectifier, transformer and bridge circuits.* A full-wave rectifier circuit requires the use of two rectifier diodes and a mains transformer of which the secondary winding is provided with a centre-tap. In the full-wave rectifier circuit of Fig. 2.8a, the secondary windings 20 V, 0, 20 V.

Observe with the cro, the voltage waveform across the 2 kΩ resistor and measure the peak voltage, V_{pk}, also measure with a suitable moving-coil voltmeter the average voltage V_{av} and establish that

$$V_{av} = 2V_{pk}/\pi$$

Note that the peak output voltage V_{pk} corresponds to that across only half of the transformer secondary winding.

If the bridge circuit (Fig. 2.8b) is used instead, the full transformer secondary voltage is available but four diodes are needed. The four diodes involve extra cost and a slightly increased voltage drop across the rectifying element, but this voltage drop is so small that generally it is of no consequence. Bridge rectifier units with four output terminals are commonly commercially available and may have built-in heat sinks so as to be capable of supplying large currents.

2.3.4 *Full-wave rectifier with smoothing capacitor.* Fig. 2.8c shows a full-wave rectifier with a centre-tapped transformer secondary, two diodes and a reservoir smoothing capacitor C, which has a value of, say, 500 μF. This form of power supply is widely used.

An experiment on this popular type of power supply is to measure the output voltage across the resistor R_L set at 3 kΩ then vary R_L between 3 kΩ and 10 kΩ and record the peak-to-peak ripple voltage for each value. Repetition of the calculations of the type shown in section 2.3.2 with, in this case, a discharge time of 0.01 s, is of interest.

Note that if the capacitance of the reservoir

capacitor is increased, the ripple voltage is smaller. Is it therefore good policy to utilise the largest value of C that is obtainable?

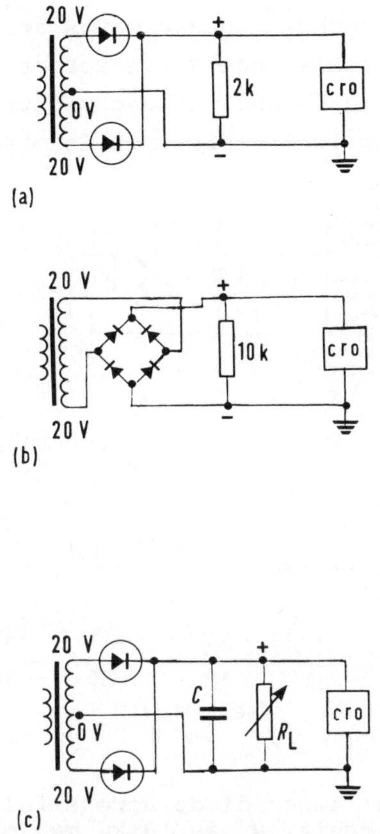

Fig. 2.8 (a) Full-wave rectification: a circuit based on the use of a centre-tapped secondary winding and two diodes; (b) a bridge circuit; (c) power supply provided by a full-wave rectifier circuit with capacitor smoothing.

2.3.5. *Zener diode across a full-wave rectifier circuit.* In Fig. 2.9a, the voltage output from a full-wave rectifier circuit is connected across a Zener or voltage-regulator diode with a 1 kΩ resistor in series. It is very important to limit the current through a Zener diode by connecting a resistance in series with it, otherwise the diode p-n junction will become over-heated and ruined.

<u>Once voltage exceeds V_Z, current tends to infinity</u>

With the load resistance R_L = 10 kΩ, the output voltage waveform is observed (Fig. 2.9b) by means of a cro. Note the production of roughly square waves at a frequency of 100 Hz. The load resistance R_L is set at various values between 200 Ω and 10 kΩ and, at each setting the voltage waveform across the load resistance is observed.

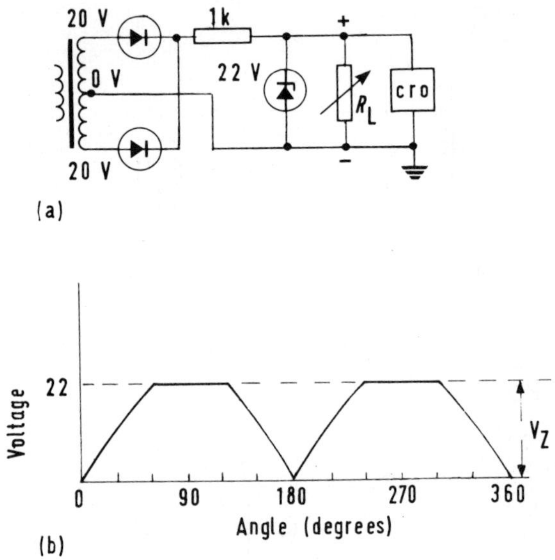

Fig. 2.9 (a) Zener diode across full-wave rectifier circuit, R_L is 10 kΩ maximum; (b) output voltage of roughly square waveform obtained.

2.3.6 *Zener diode stabilization.* In the circuit of Fig. 2.10a the full-wave rectifier voltage output is smoothed by a reservoir capacitor *C* and there is also across *C* a Zener diode with a 1 kΩ resistor in series.

Examination by a cro of the ripple of the output voltage as R_L is varied between 3 kΩ and 10 kΩ shows the effectiveness of this simple means of stabilization of voltage. Further improvement of this supply voltage stabilization is obtained by the use of two Zener diodes in parallel (each with a 1 kΩ series protective resistor) provided the second Zener has a lower voltage rating than the first (Fig. 2.10b).

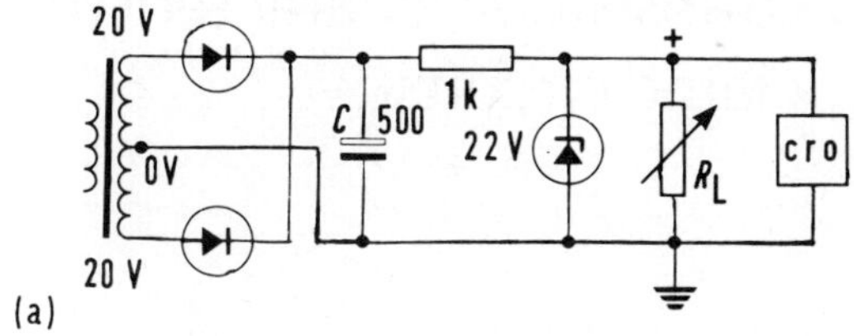

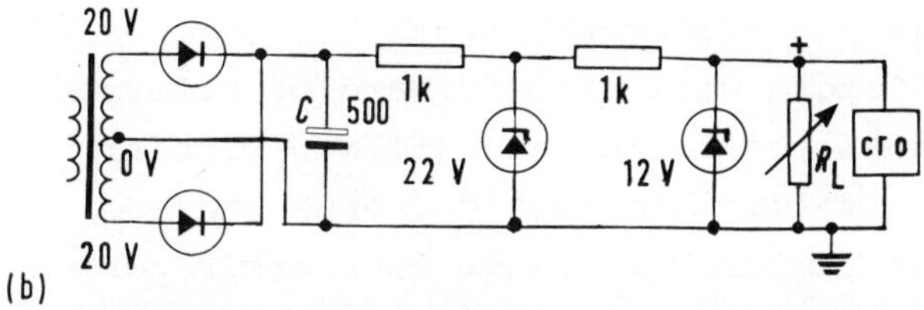

Fig. 2.10 (a) Full-wave rectification with reservoir capacitor and Zener diode stabilization; (b) improved stabilization utilizing two Zener diodes in parallel. R_L is 10 kΩ maximum in both these circuits (a) and (b)

2.3.7 *Question.* If the power rating of the Zener diodes is 1 W, calculate the maximum allowable current through (a) the 22 V Zener diode and (b) the 12 V Zener diode in Fig. 2.10b.

2.3.8 *Further investigations.* Connect up a circuit similar to that of Fig. 2.10a to provide a stable output voltage of 12 V but with an output current which may be varied between 0 and 20 mA.

CHAPTER THREE

Bipolar junction transistors: characteristics and simple associated circuits

3.1 *Bipolar junction transistors*

These transistors will henceforward be referred to as bipolar transistors. Because silicon bipolar transistors exhibit much smaller leakage currents and can be operated over a wider temperature range than their germanium counterparts, they are used almost exclusively. A convenient silicon n-p-n bipolar transistor for experiments is the BC 107.

3.2 *Characteristics of an n-p-n transistor in common-base (CB) connection*

Before beginning the experiments, the circuit (Fig. 3.1) should be examined to ensure that the polarities marked are in accordance with the principles of known bipolar transistor action. Note that in CB connection the input (steady voltage in the present case) is applied between the emitter and base whereas the output is between the collector and base, ie the base is common to both input and output circuits.

To determine the common-base input characteristics, set the collector-base pd V_{CB} to zero and measure the emitter current I_E for various values of the emitter-base pd V_{EB} up to 0.7V; the emitter current should not exceed 10 mA. Repeat the measurements of these quantities with V_{CB} constant at 2 V and then 4 V. Plot a graph of V_{EB}

against I_E with V_{CB} as parameter. The input resistance of the transistor h_{ib} (where i refers to input and b to common-base connection), is given by dV_{BE}/dI_E the gradient of the curve, taken in the case of the BC107 at I_E = 4 mA. Typical input characteristics are shown in Fig, 3.2a.

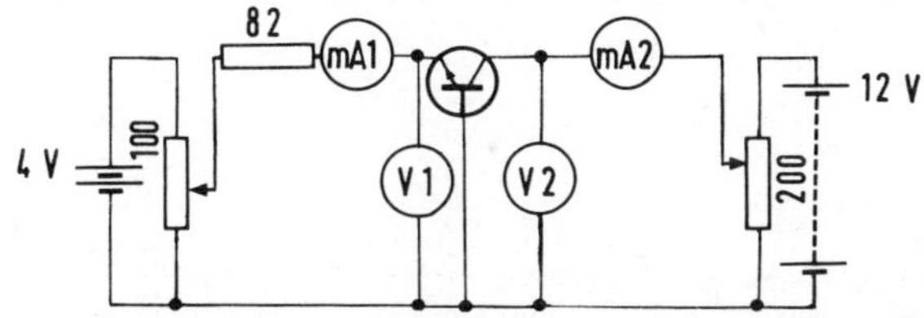

Fig. 3.1 To obtain the characteristics of an n-p-n transistor in common-base (CB) connection. Milliammeters mA1 and mA2 each have a fsd of 10 mA: mA1 measures I_E; mA2 measures I_C. V1 and V2 are both high resistance voltmeters (preferably dvms): V1 has a fsd of 1V and measures V_{EB}; V2 has a fsd of 10 V and measures V_{CB}.

3.2.1 *Output characteristics and output conductance* h_{ob}. With the circuit of Fig. 3.1, set I_E at 2 mA and maintain it constant as V_{CB} is increased in steps from 0 to 10 V. At each setting of V_{CB}, reccrd the collector current I_C. Repeat these readings for constant values of I_E at 4 mA, 6 mA, 8 mA, 10 mA. Plot a graph of I_C against V_{CB}: a typical set of output characteristics is shown in Fig. 3.2b. The output conductance h_{ob} (where o refers to output) is given by dI_C/dV_{CB} and quoted for a given value of I_E and mean value of V_{CB} expressed in siemen; the reciprocal of h_{ob} is the <u>output resistance</u> in ohms of the transistor.

The output characteristics appear to be straight lines equally spaced and all parallel to the V_{CB} - axis: it is impossible to obtain an output resistance from these curves - it suffices to show that the output resistance is very large. This very high output resistance of the bipolar transistor in CB connection is useful when it forms the basis of a constant current source (Section 3.8.1).

3.2.2 *The current gain.* This is dI_C/dI_E, and is denoted by h_{fb} and quoted for a given value of V_{CB} and mean value

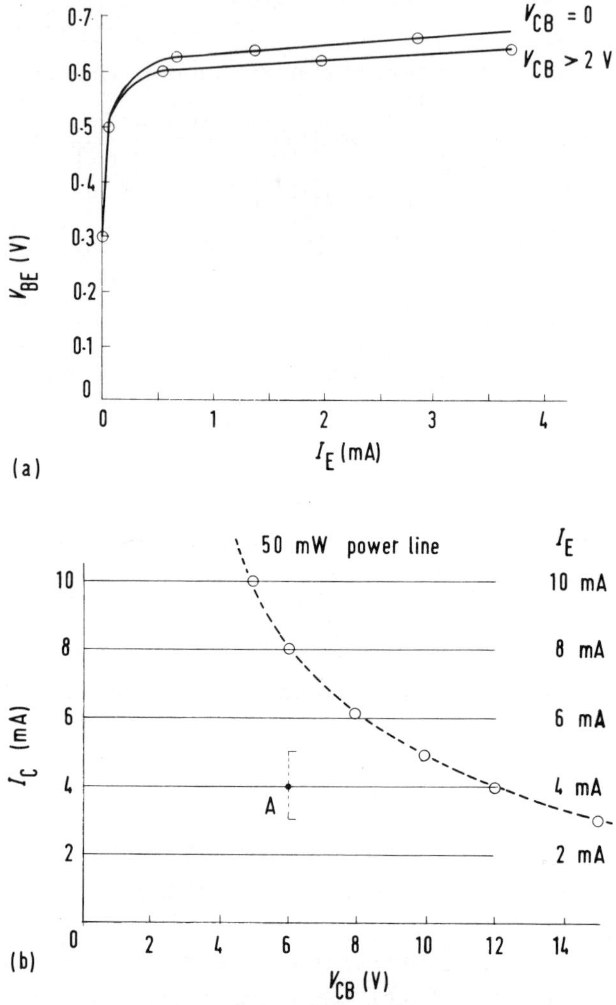

Fig. 3.2 For a silicon n-p-n bipolar transistor in CB connection: a) typical input characteristics; b) typical output characteristics.

of I_E; it may be estimated from the output characteristics. For example, at the point A (Fig. 3.2b) where I_C = 4 mA and V_{CB} = 6 V, a change of I_E by 2 mA causes a change of I_C by approximately 3 mA. Hence

$$h_{fb} \simeq 1$$

Values of I_C may be plotted against corresponding values of I_E for V_{CB} = 6 V; the curve is a straight line of gradient 1.

3.2.3 *Power considerations.* On the output characteristics (Fig. 3.2b) a curve is drawn corresponding to a power dissipation of 50 mW at the collector-base junction. As this junction is reverse-biased, the power dissipated is simply the product of I_C and V_{CB}. Any portion of a characteristic obtained by experiment must lie well clear (to the left in Fig. 3.2b) of the maximum power dissipation curve. When plotting a characteristic, one should always find the power rating of the transistor from the manufacturer's literature; it is rarely necessary to exceed a dissipation of 100 mW.

Note: The currents in the collector, emitter and base are related by the equation

$$I_C + I_B = I_E$$

whatever the circuit configuration; hence I_C is always less than I_E, but as I_B is small compared with the other currents (generally $I_B \simeq 0.005\ I_E$), $I_C \simeq I_E$

3.2.4 *More accurate value of* h_{ob}. Manufacturers normally quote the *h* parameters for the common-emitter connection where

h_{ie} is the input resistance
h_{oe} is the output conductance
h_{fe} is the current gain.

These *h* parameters can readily be determined by graphical methods for the common-emitter connection (p.31) and those for the CB connection can be calculated from these. This is particularly useful in determining h_{ob} because

$$h_{ob} = h_{oe}/(1 + h_{fe})$$

For example, a silicon bipolar transistor has h_{fe} = 150, and $h_{oe} = 30 \times 10^{-6}$ siemen. Hence

$$h_{ob} = h_{oe}/(1 + h_{fe}) \simeq 30 \times 10^{-6}/150 = 2 \times 10^{-7} \text{siemen.}$$

The output resistance ($1/h_{ob}$) is therefore 5 MΩ, which could not have heen determined from the output characteristics of Fig. 3.2b.

3.3 *Characteristics of an n-p-n bipolar transistor in common-emitter (CE) connection*

The circuit of Fig. 3.3 is used. Set $V_{CE} = 0$ and record the base current I_B corresponding to various values of V_{BE}, where the base current should not exceed 60 μA. Repeat these readings with $V_{CE} = 2$ V and then 4 V. Plot V_{BE} against I_B and determine the gradient (dV_{BE}/dI_B) of the curve at $I_B = 20$ μA. This gradient gives the input resistance of the transistor, denoted by h_{ie}.

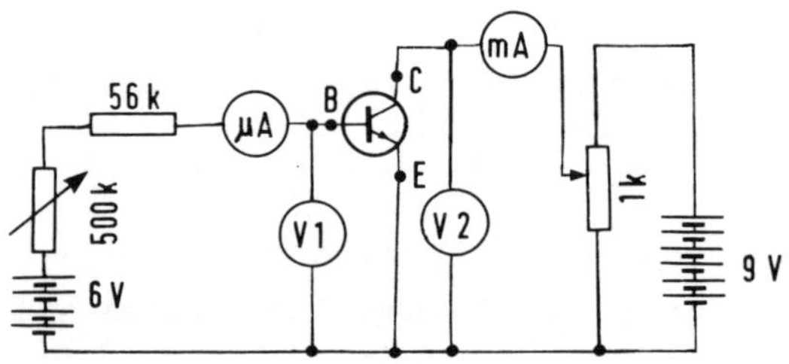

Fig. 3.3 To obtain the characteristics of an n-p-n bipolar transistor in CE connection. μA is a microammeter with a fsd of 100 μA and measures I_B; mA is a milliammeter with a fsd of 10 mA and measures I_C. V1 and V2 are both high resistance voltmeters (preferably dvms): V1 has a fsd of 1 V and measures V_{BE}; V2 has a fsd of 10 V and measures V_{CE}.

3.3.1 *Output characteristics and the output conductance* h_{oe}.

With the circuit of Fig. 3.3 set the base current I_B at 10 μA and record the collector current I_C for various settings of the collector-emitter pd V_{CE}; do not allow I_C to exceed 10 mA. Repeat these readings with $I_B = 20$ μA, 30 μA, 40 μA and 50 μA. Plot the output characteristics

(I_C against V_{CE}) for these constant values of I_B. At the point I_C = 4 mA, V_{CE} = 6 V, determine the gradient (dI_C/dV_{BE}) of the output characteristics. This gradient will have the dimensions of conductance: it is the output conductance of the transistor denoted by h_{oe}. Typical output characteristics are shown in Fig. 3.4.

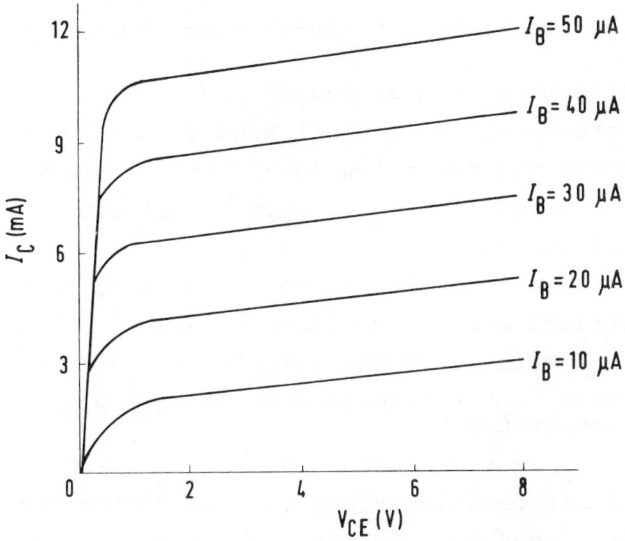

Fig. 3.4 Typical output characteristics of an n-p-n transistor in CE connection

3.3.2 *The current gain* h_{fe}. At the point I_C = mA, V_{CE} = 6 V on the output characteristics (Fig. 3.4) determine

$$h_{fe} = dI_C/dI_B$$

If a change in I_C of 2 mA is produced by a change in I_B of 15 μA, then

$$h_{fe} = 2 \times 10^{-3}/(15 \times 10^{-6}) \simeq 135 .$$

3.3.3 *Power dissipation.* On the output characteristics, plot the 50 mW power curve.

Construct a table to compare the values for the given bipolar transistor of the input resistance, output resistance and current gain for the common-emitter connection with those obtained for the common-base connection, ie compare h_{ie} with h_{ib}, $1/h_{oe}$ with $1/h_{ob}$ and h_{fe} with h_{fb}. Check the relationship

$$h_{ib} = h_{ie}/(1 + h_{fe}) \qquad (3.1)$$

The results obtained should show that the input resistance in the common-emitter connection is greater than in common-base whereas the output resistance is lower.

3.4 *A bipolar transistor tester*

It is of great help if a small unit is available which can check quickly whether a low-power bipolar transistor or a diode is functional or not. Such a unit can be readily constructed and used.

For a bipolar transistor, the answers to the following three questions are of immediate value.

(i) Are both junctions intact?

(ii) What is the leakage current I_{CEO} at room temperature?

(iii) What is the current gain h_{FE}?

The circuit configurations used to define the leakage currents I_{CBO} and I_{CEO} for an n-p-n bipolar transistor are shown in Figure 3.5.

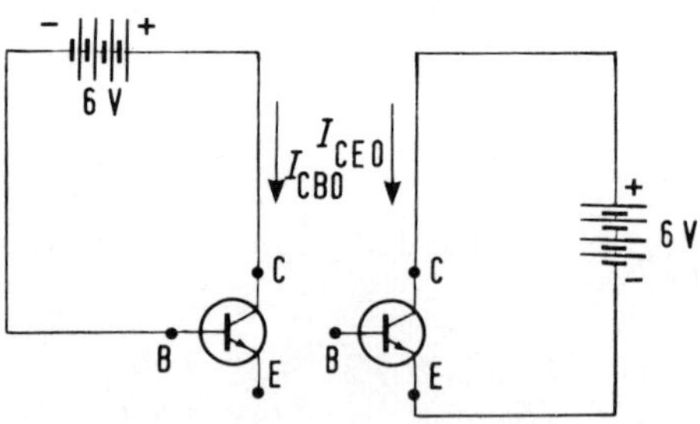

Fig. 3.5 Circuit configurations used to define I_{CBO} and I_{CEO} for an n-p-n bipolar transistor.

It may be shown that

$$I_{CEO} \simeq h_{FE}\, I_{CBO}$$

Manufacturers usually quote the values of I_{CBO} and h_{FE} at 20°C so that I_{CEO} may be calculated.

If, for example, for a low-power germanium bipolar transistor at 20°C I_{CBO} is 3 µA and h_{FE} = 100, I_{CEO} = 300 µA. This leakage current doubles approximately (in the case of germanium) for every 10°C rise in temperature. For a silicon bipolar transistor of similar power handling capability, I_{CEO} may be 1 µA and this changes insignificantly for a 10°C rise in temperature.

The test circuit of Fig. 3.6 is arranged so that the transistor is not harmed even if it is plugged in the wrong way round. Furthermore the meter is protected and the power dissipated in the transistor can never damage the junctions.

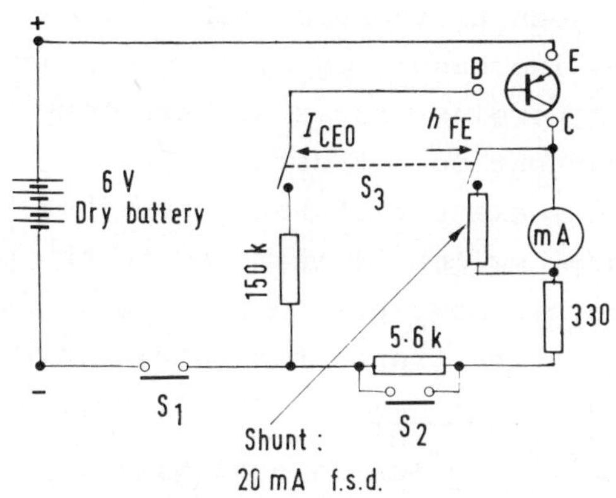

Fig. 3.6 The bipolar transistor tester. mA is a milliammeter with a centre zero and a scale reading 2-0-2 mA.

(a) Examine the features of this transistor tester circuit (Fig. 3.6), where the battery is connected to test a p-n-p bipolar transistor. For an n-p-n device, the battery connections would need to be reversed. S_1 and S_2 are press-button switches which are closed when depressed.

Assume that a p-n-p transistor is inserted in the socket and that, with the switch S_3 in the position I_{CEO}, if switch S_1 is pressed, the meter reading is 1 mA approximately: the transistor is useless because E and C are virtually shorted due to the fact that the collector-base junction is damaged. (Note that the total resistance in the circuit = 5.6 k Ω + 330 Ω ≃ 6k Ω, so the current = 6/6000 = 1 mA).

However, if the collector-base junction of the transistor is intact, only a small leakage current will be recorded when S_1 is closed and the true leakage current can also be measured when S_2 is closed because the 6V battery is connected as indicated in Fig. 3.6. One would expect I_{CEO} for a germanium bipolar transistor to be less than 400 μA whereas for a silicon transistor it would be undetectable. While S_1 and S_2 are depressed and I_{CEO} is being recorded, the leakage current of the germanium transistor can be seen to increase markedly as the junction temperature is increased. eg by simply placing a finger on the transistor housing; for a satisfactory silicon device, the leakage current is not detectable.

(b) If the leakage test has shown that the transistor is satisfactory, switch S_3 is moved to the position marked h_{FE}. The meter is now shunted to give an fsd of 20 mA and current flows in the transistor base. This base current I_b is given by $\frac{6}{150 \times 10^3}$ A = 40 μA, because the pd across the emitter-base junction is negligible. If S_1 and S_2 are pressed, the meter reads the collector current: if this is 10 mA, the current gain of the transistor, $h_{FE} = I_c/I_b$ = 10 mA/40 μA = 250.

(c) If it is required to test a diode or perhaps to identify its electrodes it may be plugged between terminals E and C. When switch S_1 is pressed the meter reading will be 1 mA approximately if the diode is forward-biased and zero if reverse-biased. A reading of 1 mA obtained for each direction (obtained by reversing either the diode or the battery connections) indicates that the junction is damaged and the diode is useless. The meter should not be shunted when testing diodes, ie switch S_3 should be in the position I_{CEO}.

3.5 *Further investigation*

Mount a germanium bipolar transistor on a piece of cork and float it on the surface of maize oil with the transistor completely immersed. With a 4.5V dry battery connected between the collector and the emitter as shown in Fig. 3.5, measure the leakage current I_{CEO} at a number of temperatures between 0°C and 60°C. Plot a graph of leakage current against temperature and determine dI_{CEO}/dT at 30°C.

3.6 *Voltage stabilizing circuits: general information, the use of bipolar transistors*

The output voltage V_O from a power supply may be subject to a change ΔV_O. This change can be brought about by changes in the input voltage V_i by ΔV_i, and the ambient temperature T by ΔT. Thus, the performance of any voltage stabilizer can be expressed by the equation

$$\Delta V_O = S_V \, \Delta V_i + R_O \, \Delta I_L + S_T \, \Delta T \qquad (3.2)$$

where the coefficients S_V, R_O and S_T are defined as follows.

stabilization factor $\quad S_V = \left| \dfrac{\partial V_O}{\partial V_i} \right|_{I_L, T}$

output resistance $\quad R_O = \left| \dfrac{\partial V_O}{\partial I_L} \right|_{V_i, T}$

temperature coefficient $\quad S_T = \left| \dfrac{\partial V_O}{\partial T} \right|_{I_L, V_i}$

The object of the design of a voltage stabilizing circuit is to produce a constant V_O (ie $\Delta V_O \to 0$) in spite of changes in V_i, I_L and T. To simplify, it is assumed that the ambient temperature T is constant and hence only S_V and R_O are significant. In any case, to ensure that the circuit is relatively insensitive to temperature change (ie S_T is small) silicon transistors are invariably used, the power demanded from the circuit is restricted and the output transistor is mounted on a heat sink.

In a well-designed voltage stabilizing circuit S_V and R_O are both as small as possible. An input voltage change ΔV_i of a dc power supply based on the ac mains may be ripple due to inadequate filtering at the rectifier output (eg the smoothing capacitor used has too small a capacitance). The ability of the circuit to prevent these variations ΔV_i from affecting V_O is governed by the magnitude of S_V. The output resistance R_O determines how constant the output voltage remains when the load current changes.

3.6.1 *Emitter follower voltage stabilizer.* If the load current demanded is greater than can be provided by a simple Zener diode stabilized voltage supply circuit (section 2.3.6) or if the load resistance is likely to change considerably, a voltage stabilizer which makes use of a series control transistor is preferable (in Fig. 3.7, two bipolars connected as emitter-followers are used). The reference voltage is still provided by a Zener diode but it is now used in a bipolar transistor emitter-follower circuit so that the current through the diode is almost constant.

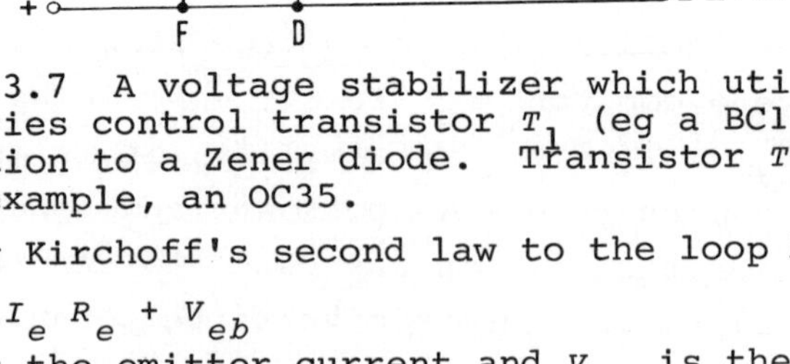

Fig. 3.7 A voltage stabilizer which utilizes a series control transistor T_1 (eg a BC187) in addition to a Zener diode. Transistor T_2 is, for example, an OC35.

Applying Kirchoff's second law to the loop AFDEBA gives

$$V_Z = I_e R_e + V_{eb}$$

where I_e is the emitter current and V_{eb} is the equivalent diode voltage across the emitter-base junction which is constant at approximately 0.5 V for a silicon p-n junction at constant temperature. Hence

$$V_Z \simeq I_e R_e$$

so that whatever voltage is applied between the base and the emitter of the transistor will appear across the resistance R_e in the emitter lead. Depending on the setting of the tapping point on the 1 kΩ potentiometer used for R_e in Fig. 3.7 a chosen fraction of this voltage is applied between the base and the emitter of the output transistor T_2. This chosen voltage ($\leq V_Z$) must appear across any load resistance R_L and should remain constant irrespective of a change ΔR_L in R_L or a change ΔV_i in V_i.

To examine the circuit of Fig. 3.7 experimentally, apply 25 V across its input terminals (ie V_i = 25 V) where T_1 is type BC187 and T_2 an OC35. Measure the voltage V_Z across the Zener diode and the voltage across the 1k Ω potentiometer in the emitter lead of T_1. Set R_L = 50k Ω and adjust the 1k Ω potentiometer to give the maximum output voltage, V_{OMAX}. Record the output voltage V_{OMAX} and the output current I_O at each setting of R_L, where I_O should not exceed 200 mA. Repeat with the potentiometer R_E set to give an output voltage V_O of 10 V and again with V_O at 5 V. Plot curves of V_O against I_O and determine the gradient of each curve; this gradient is the output resistance R_O of the stabilizing circuit given by

$$R_O = \left| \frac{\partial V_O}{\partial I_O} \right|_{V_i, T}$$

Again set V_i = 25 V and R_L = 50k Ω and the output voltage at its maximum value V_{OMAX} making use of the 1k Ω potentiometer. Adjust R_L to give a current I_O of 100 mA. Record the output voltage V_O. Decrease the input voltage V_i in steps over the range between 25 V and 16 V, keeping the current I_O constant at 100 mA. Measure V_{OMAX} at each setting of V_i.

Reset the input voltage V_i to 25 V, R_L to 50 kΩ and adjust the output voltage V_O to 10 V. With the output current constant at 100 mA, record values of V_O as V_i is decreased in steps from 25 V to 16 V. Plot curves of V_O against V_i. The gradient

$$\left|\frac{\partial V_O}{\partial V_i}\right|_{I_O, T} = S_V$$

of each curve gives the stabilization factor S_V which, of course, should be as small as possible.

3.7 *Constant current sources: introduction*

A constant current source is designed to have a very high (ideally infinite) output resistance where the output resistance is the gradient ($\partial V/\partial I$) of the output characteristic. The desirable consequence of a high output resistance is that any change in the load resistance has little effect on the current which therefore remains nearly constant at its selected value.

Variations in temperature and in input voltage can also affect the current flowing and the output resistance; the influence of these variables must be minimized.

3.7.1 *A constant current source based on a bipolar junction transistor.*

In common-base connection a bipolar transistor has a very high output resistance (Section 3.2.1); this feature is utilized in the simple constant current source circuit of Fig. 3.8a, which can provide a constant current of up to 10 mA. In this circuit, the effects of temperature change are reduced by

(i) using a silicon transistor

(ii) using two 5.6 V Zener diodes in series, where the temperature coefficient (dV_z/dT) of these diodes is nearly zero,

(iii) limiting the current to 10 mA, giving negligible heating at the reverse-biased junction; also the transistor is mounted on a small heat sink.

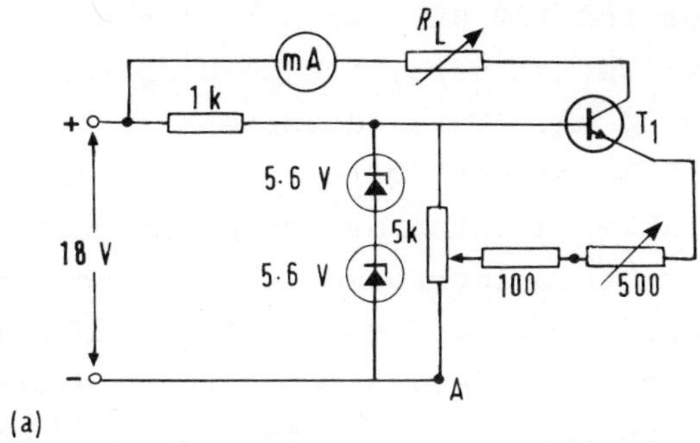

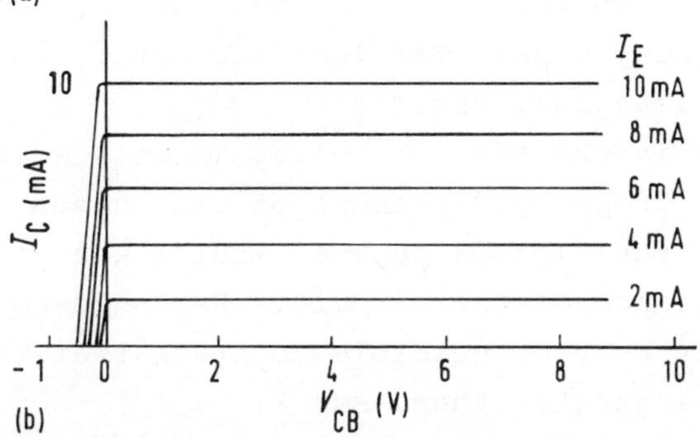

Fig. 3.8 (a) Constant current source based on a bipolar transistor T_1 (eg 2N2219) in CB connection. (b) Typical output characteristics of a bipolar transistor in CB connection.

Fig. 3.8b shows typical output characteristics for a bipolar junction transistor in common-base connection. Note that the collector current I_C is constant and independent of the collector-base voltage V_{CB} only if the emitter current is constant. To achieve this and yet allow the constant value of I_E to be selected, the emitter-base voltage V_{EB} is obtained from a Zener diode stabilized supply in the circuit (Fig. 3.8a). The collector-base voltage V_{CB} with R_L = 0 is the voltage across the 1 kΩ resistor, ie approximately 18 - 11.2 = 6.8 V. Hence the load current should remain constant until the voltage drop across the load resistor approaches 6.8.

An 18 V dc supply is used in the experiment on the circuit of Fig. 3.8a, the milliammeter, mA has a fsd of 10 mA or later in the experiment is replaced by a microammeter of fsd 100 μA. With R_L (obtained from a resistance box) set at zero, the load current I_L is adjusted to be 10 mA. This adjustment is obtained by utilizing the 5 kΩ potentiometer as a coarse control and the 500 Ω variable resistor as a fine control. The load resistance R_L is then increased in steps and the load current I_L recorded at each value of R_L; this is continued until the load current begins to decrease appreciably. This procedure is repeated with initial load currents of 5 mA, 1 mA and 100 μA. The load current I_L is plotted against the load resistance R_L.

(a) What is the maximum voltage across R_L before the load current drops by 3 percent of its chosen value?

(b) From the plotted curves, indicate a value for the output resistance of the circuit. Having used moving-coil meters it may only be possible to state that the output resistance is greater than, say X.

3.7.2 *Further investigations*

(a) Use this source to maintain a constant current of, say 10 mA through a negative temperature coefficient thermistor, then measure the voltage across the thermistor as its temperature is increased in steps to 200°C by immersing it in a heated beaker of oil of which the temperature is recorded by a mercury-in-glass thermometer.

(b) An important equation concerned with the electrical conductivity of an intrinsic specimen of germanium is

$$\sigma = \sigma_0 \exp(-E_G/2kT)$$

where E_G is the energy gap between the top of the valence band and the bottom of the conduction band, k is the Boltzmann constant and T is the absolute temperature.

Again with a constant current maintained by this source through a specimen of germanium*, measure the voltage

*See also Close and Yarwood, An Introduction to Semi-Conductors (1971), Heinemann, London, p.420.

across it at a number of temperatures. From the slope of a plot of $log_{10}V$ against $(1/T)$ a value for E_G may be obtained.

(c) Use this source to maintain a constant current through a Hall probe being utilized for determining the distribution of magnetic flux density due to a powerful magnet.

3.8 *Amplifiers: use of bipolar transistors*

In electronics an amplifier is equipment which provides at its output terminals a magnified replica of the signal supplied to its input terminals.

In general, the input signal is varying and, in many (indeed, the majority of) cases, it is alternating and is either in the form of a small alternating current or a small alternating voltage. There is also an important class of amplifiers in which the input signal is steady, often a small direct current, and where the purpose of the amplifier is to provide at its output a much larger direct current.

When the input signal is alternating, the current or voltage concerned is usually small and controls (via the amplifier) a large output power which can be either in the form of a large alternating voltage across a high resistance or a large alternating current through a low resistance.

In the perfect amplifier every feature of the input signal is present at the output: the waveform output power (voltage or current) is precisely the same as the input signal waveform. Any change in the waveform introduced by the amplifier between the input and the output is termed *distortion* - some distortion (albeit often insignificant in amount) is inevitable. Furthermore, the amplifier may itself introduce features at the output which are not present in the input signal. Such detail added by the amplifier is termed *noise*.

In any electronic amplifier, the power available at the output stage is provided by the dc power supply: the input signal thus controls the amplifier so that it converts some of the dc power into ac power of specific waveform.

Bipolar transistors can be used in voltage amplifier stages in one of the three basic connections: common-base (CB), common-collector (CC) and common-emitter (CE). The last of these (CE) is used so much more frequently than the others that it is the only one of the three examined experimentally here. Its name derives from the fact that the emitter terminal is common to both the input and output circuits (Fig. 3.9). In this amplifier stage a constant direct current passes continuously through the transistor; this direct current fixes the operating point on the characteristic. The small alternating current signal to be amplified is superimposed on this direct current, this signal being applied between the base and the emitter of the bipolar transistor; the capacitor C_1 is inserted to isolate the signal source from the steady power supply voltage V_{CC}. If the operating point has been chosen correctly and if the input signal is of small amplitude, at the output terminals will appear the output signal which is an amplified, inverted replica of the input signal, with the input and output signal 180° out-of-phase.

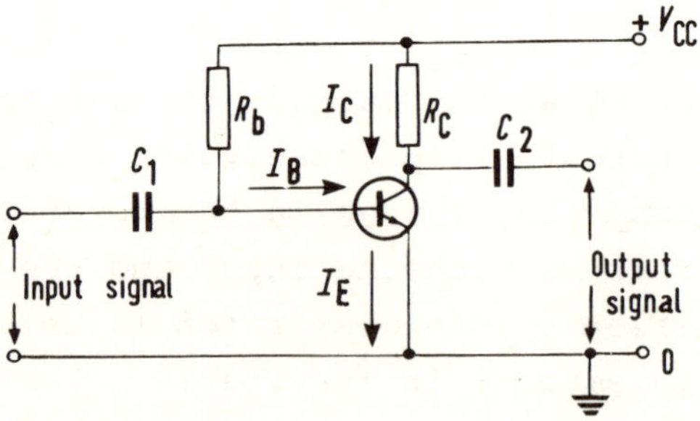

Fig. 3.9 A bipolar transistor common-emitter amplifier with simple fixed bias.

In Fig. 3.10, again showing a CE amplifier, the dc bias arrangements are provided by means of a potential divider network (R_1 and R_2) and an emitter resistor R_E. These components determine the operating point and the steady direct current through the transistor. This type of biasing, sometimes called 'base voltage biasing' is thermally more stable and is much more frequently used than the simple fixed bias arrangement shown in Fig. 3.9.

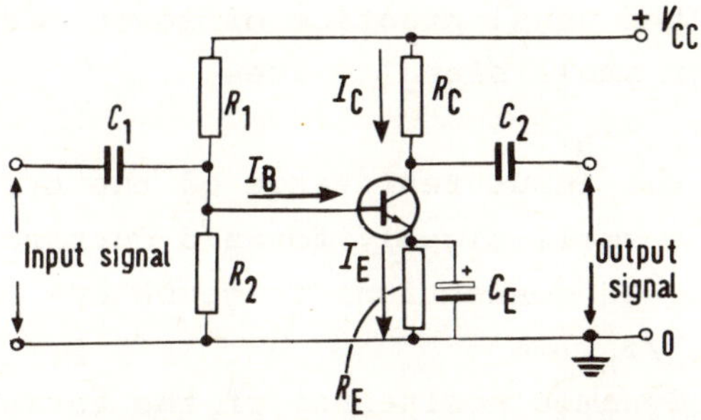

Fig. 3.10 A bipolar common-emitter amplifier with potential divider and emitter biasing resistor.

3.8.1 *A common-emitter (CE) bipolar transistor single-stage amplifier.* To simplify calculation of the ac behaviour of such an amplifier, it is useful to consider an equivalent circuit; in such an equivalent circuit, information about the dc bias is omitted: it is assumed that the device is correctly biased, with the emitter-base junction forward-biased and the collector-base junction reverse-biased. The transistor (an active, non-linear device) may be replaced by resistive elements which are passive and a constant current source, provided that calculations are restricted to small variations about the operating point over which limited region the transistor characteristics may be assumed to be linear. Capacitance is omitted in an equivalent circuit unless one needs to assess its effect over certain regions. Thus a bipolar transistor in CE connection can be represented by the simple, small-single equivalent circuit of Fig. 3.11.

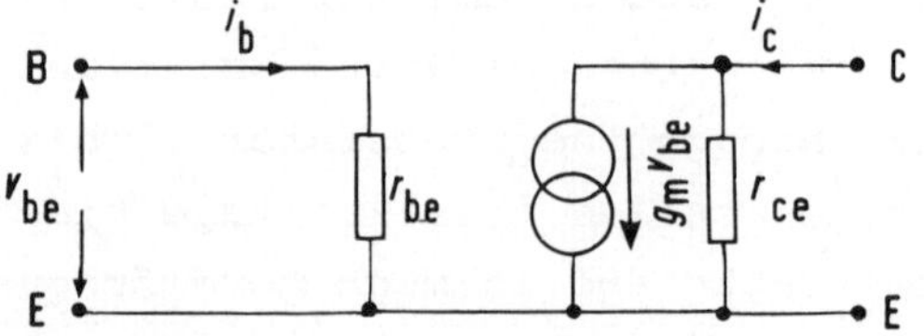

Fig. 3.11 A simple equivalent circuit for a bipolar transistor in CE connection

Following the usual practice of lower case letters and subscripts for small signal values,

$$r_{be} = v_{be}/i_b \tag{3.3}$$

where r_{be} is the input resistance of the transistor.

As h_{fe} (the small-signal, forward current gain of the transistor in CE connection) is given by

$$h_{fe} = i_c/i_b \tag{3.4}$$

and r_e (the dynamic resistance of the forward-biased base-emitter junction of the transistor) is given by

$$r_e = v_{be}/i_c \tag{3.5}$$

it follows that equation (3.3) may be expressed in the form

$$r_{be} = h_{fe}\ r_e\ . \tag{3.6}$$

As

$$r_e = (0.026/I_E)\ \Omega$$

where I_E is the steady emitter current in amperes,

$$r_{be} = (h_{fe}\ 0.026/I_E)\ \Omega \tag{3.7}$$

In Fig. 3.11, the current generator provides a current i_c of magnitude decided by the equation

$$i_c = g_m\ v_{be}$$

Therefore

$$g_m = i_c/v_{be} = h_{fe}\ i_b/v_{be} = h_{fe}/r_{be} \tag{3.8}$$

(see equations 3.4, 3.5 and 3.6)

From equation (3.7) it follows that

$$g_m = h_{fe}/(h_{fe}\ 0.026/I_E) = 38\ I_E \tag{3.9}$$

In the circuit diagram of Fig. 3.11, the resistance r_{ce} is usually large enough to be ignored: thus, for a small-signal silicon planar transistor it is about 100 kΩ at I_E = 1 mA and r_{ce} = const/I_E so that the current through r_{ce} is negligibly small.

A simple CE amplifier stage is shown in Fig. 3.12a; it has the equivalent circuit of Fig. 3.12b. Simple calculation from the values given in the former circuit (Fig. 3.12a) shows that, relative to earth, the base potential is about + 3.1V so that the emitter potential is about + 2.5V. Hence, the steady current through the resistance in the emitter lead must be

$$I_E = \frac{2.5}{5\times10^3}\ \text{A} = 0.5\ \text{mA}$$

From equation (3.9) it follows that

$$g_m = 38I_E = 19\ \text{mA V}^{-1} = 0.019\ \text{A V}^{-1}$$

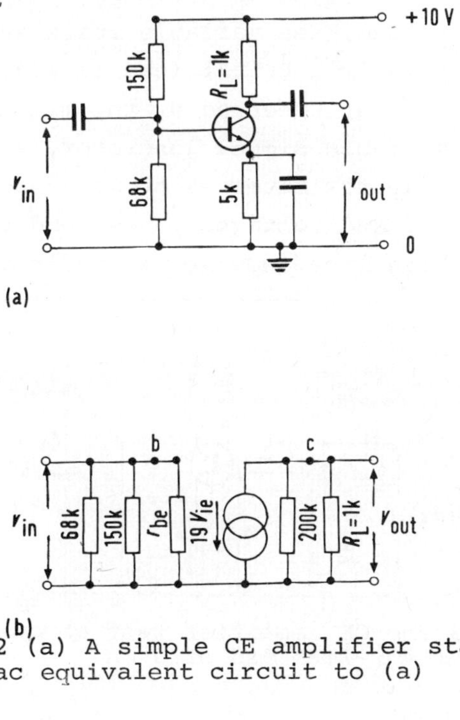

Fig. 3.12 (a) A simple CE amplifier stage.
(b) The ac equivalent circuit to (a)

As r_{ce} (see Fig. 3.11) can be ignored, the voltage gain A_V is given by

$$A_V = \frac{\text{output voltage}}{\text{input voltage}} = \frac{g_m v_{be} R_L}{v_{be}} = g_m R_L \qquad (3.10)$$

which, substituting $g_m = 0.019$ A V^{-1} and $R_L = 1000$ gives $A_V = 19$ whereas for $R_L = 100$, $A_V = 1.9$.

If this simple analysis is correct, all small signal transistors used in the circuit of Fig. 3.12a would give a voltage gain of about 19, for the same value of I_E, irrespective of their current gains (h_{fe}). It would appear therefore that the current gain of the transistor affects the input resistance of the CE amplifier but does not affect the voltage gain.

The experiment designed to test the validity of these relationships is based on the common-emitter single-stage amplifier of Fig. 3.13 which is operated with a stabilised power supply of 10 V, and in which $R_L = 100\ \Omega$. With no input signal applied, the variable resistance R_{V1} in the transistor (initially a BC107) lead is adjusted so that the the collector current recorded by the milliammeter mA is 1 mA. From a suitable signal generator, a sinusoidal input signal of approximately 5 mVrms at a frequency of 1 k Hz is applied and voltages v_1, v_2 and v_3 are measured by a suitable high impedance ac millivoltmeter.

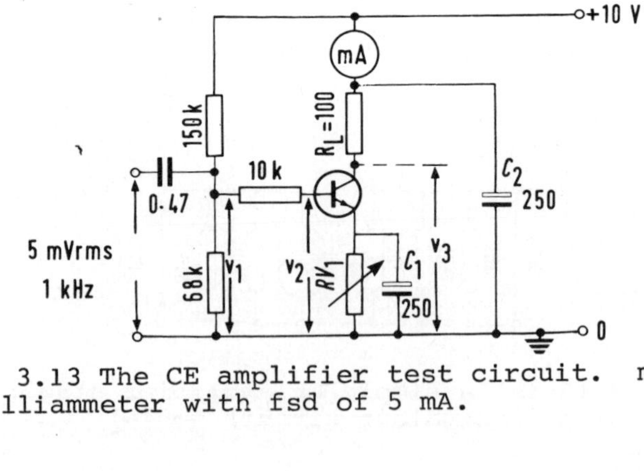

Fig. 3.13 The CE amplifier test circuit. mA is a milliammeter with fsd of 5 mA.

The 10^4 Ω resistor in the base lead to the transistor and the 10^2 Ω resistor in the collector lead should each be known to within ± 1 percent. It follows that the alternating currents i_b and i_c as well as the input resistance r_{be} can be determined because

$$i_b = (v_1 - v_2)/10^4; \; i_c = v_3/10^2 \text{ and } r_{be} = v_2/i_b$$

The readings are repeated with each of two other different n-p-n silicon bipolar transistors, eg a BC109 and then a BC171, in place of the BC107, and a table of results (as in Table 3.1) is drawn up.

Table 3.1 Experimental results in the determination of the voltage gain A_V for a CE bipolar transistor single-stage amplifier with three different transistors

Transistor *Type*	I_e mA	v_1 mV	v_2 mV	v_3 mV	Voltage gain $A_V = v_3/v_2$	Theoretical voltage gain $= 38\, I_E R_L$
BC 107	1	5	2.2	8.1	3.7	3.8
BC 109	1	5	2.9	10.6	3.7	3.8
BC 171	1	5	2.4	9.0	3.8	3.8

From such results it is seen that equations (3.9) and (3.10) are valid. In particular, it follows that the voltage gain A_V is indeed independent of the current gain h_{fe} of the bipolar transistor used, when I_E has the same value in each case.

Simultaneous display on a double beam cro of the input signal voltage v_1 and of the output voltage v_3 shows that the output is an amplified replica of the input without significant distortion and also that v_3 is 180° out of phase with v_1.

Further experiments of value with the circuit of Fig. 3.13 are

(i) With I_E = 1 mA and v_1 = 5 mV rms at 1 k Hz, disconnect the capacitor C_2 and record v_2 and v_3. Calculate the new voltage gain of the stage and compare it with the voltage gain found with C_2 connected. Explain why the voltage gain has changed.

(ii) With C_2 in circuit, I_E = 1 mA and v_1 = 5 mV rms at 1 k Hz, remove the capacitor C_1 (thus introducing negative feedback). Measure v_2 and v_3, and calculate the voltage gain of the stage. Explain why the gain has changed on this occasion. Display on a double-beam cro the voltage waveform v_2 and simultaneously the voltage across RV_1. Note that any signal across a load in the emitter lead is in-phase with the input signal.

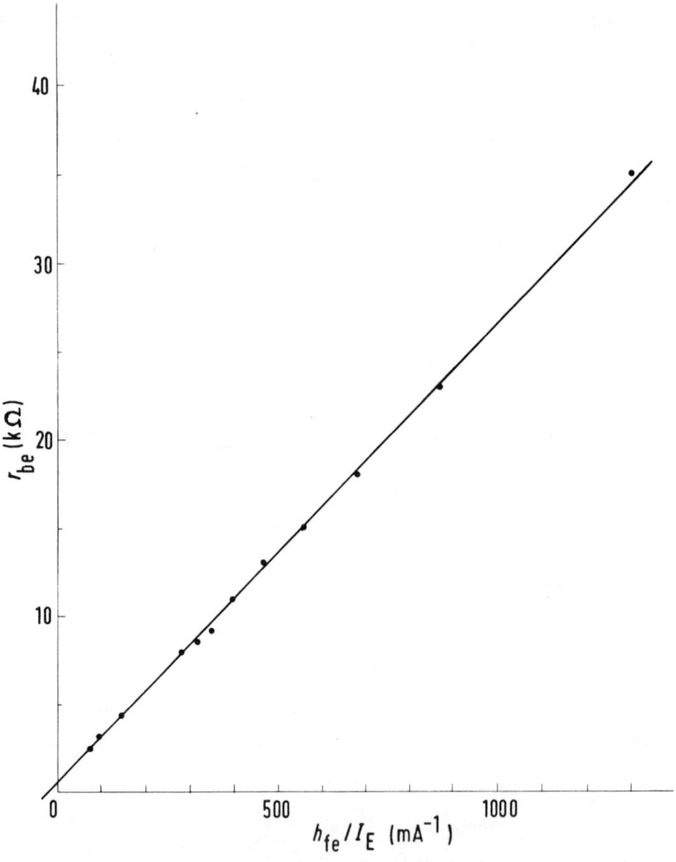

Fig. 3.14 Typical plot of results for the bipolar transistor BC107 from Table 3.2. Slope = $(35 \times 10^3)/1340$ = 26mV.

(iii) With the circuit as in Fig. 3.13, adjust RV_1 so that $I_E = 0.1$ mA. With an input signal v_1 of approximately 5 mVrms at 1 k Hz measure the alternating voltages v_1, v_2 and v_3. Increase I_E to 0.2 mA and repeat these measurements. Continue to increase I_E in steps up to 4 mA and repeat the measurements at each stage. Draw up a table of results as in Table 3.2, which includes calculated quantities.

Obtain a second table of results for a different n-p-n silicon bipolar transistor. On the same set of axes plot r_{be} in kΩ against h_{fe}/I_E in mA^{-1} for each transistor. Determine the slope of each straight line plot. Typical results are shown in Fig. 3.14.

(iv) From your table of results plot on the same set of axes g_m (in mA V^{-1}) against I_E (in mA) for each transistor, using values of I_E between 0 and 1 mA.

Determine the slope of the curves in the units V^{-1}. On the same set of axes, plot the theoretical relationship $g_m = 38\ I_E$. Excellent agreement between the experimental and theoretical plots should be found.

(v) From your table of results plot curves to show the variation in h_{fe} with I_E for each of the transistors examined.

3.9 *Sinusoidal waveform generators*

A sinusoidal waveform generator is one which provides at an instant of time t an output of magnitude y (where y is a voltage, current or power) and

$$y = y_o \sin(\omega t + \emptyset)$$

where y_o is the peak value (amplitude) of y, ω is the pulsatance ($\omega = 2\pi f$, where f is the frequency) and $\emptyset$ is a phase angle so that $y = y_o$ when $\sin(\omega t + \emptyset) = 90^o$. Usually the sinusoidal output is provided continuously, so that the output of magnitude y occupies a cycle of period $T (= 1/f)$ and the cyles are repeated, one after the other,

Table 3.2

Experimental results obtained with a CE bipolar single-stage amplifier as in Fig. 3.13

I_E mA	v_1 mV	v_2 mV	v_3 mV	$i_b = (v_1 - v_2)/10^4$ µA	$i_c = v_3/10^2$ µA	$v_{be} = v_2$ mV	$g_m = i_c/v_{be}$ mA V^{-1}	$r_{be} = v_{be}/i_b$ kΩ	$h_{fe} = i_c/i_b$	h_{fe}/I_E mA^{-1}
0.1	5	4.4	1.6	0.06	16	4.4	3.6	73	270	2700
0.2	5	3.9	2.9	0.11	29	3.9	7.6	35	260	1300
0.3	5	3.5	3.9	0.15	39	3.5	11	23	260	870
0.4	5	3.2	4.8	0.18	48	3.2	15	18	270	680
0.5	5	3.0	5.5	0.20	55	3.0	18	15	280	560
0.6	5	2.8	6.2	0.22	62	2.8	22	13	280	470
0.7	5	2.6	6.7	0.24	67	2.6	26	11	280	400
0.8	5	2.4	7.2	0.26	72	2.4	30	9.2	280	350
0.9	5	2.3	7.7	0.27	77	2.3	33	8.5	290	320
1.0	5	2.2	8.1	0.28	81	2.2	37	7.9	290	290
2.0	5	1.5	10.4	0.35	104	1.5	69	4.3	300	150
3.0	5	1.2	11.3	0.38	113	1.2	94	3.2	300	100
4.0	5	1.0	11.9	0.40	119	1.0	119	2.5	300	75

so long as the generator is switched on.

Such a sinusoidal oscillator, nominally of single frequency f and pure sine waveform output, is provided by several circuits. Three of particular interest, and capable of providing a sinusoidal output of excellent purity of waveform, are the phase-shift oscillator the crystal-controlled oscillator and the Wien bridge oscillator.

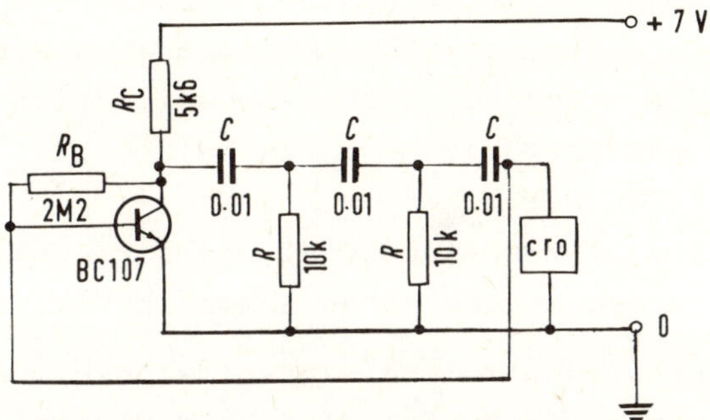

Fig. 3.15 A phase-shift sinusoidal oscillator based on a bipolar transistor (eg type BC107)

3.9.1 ***A phase-shift sinusoidal oscillator based on a bipolar transistor.*** **Fig. 3.15 is the circuit diagram of a** phase-shift sinusoidal oscillator in which a bipolar transistor amplifier is used. To ensure the necessary positive feedback from the output to the input, the collector of the bipolar transistor is connected via a capacitance resistance network to the base of the bipolar transistor. This network consists of three CR combinations where, in Fig. 3.15, the third resistance is the input resistance of the bipolar transistor.

In each of these CR arrangements, the alternating current leads by an angle α on the voltage, where

$$\alpha = \tan^{-1}(1/\omega CR)$$

Choosing three CR arrangements, each nominally the same (in fact, the three CR sections do not have to be identical because the frequency of the oscillation will be such as to ensure that the part of the feedback signal providing a phase shift of 180° will be amplified more than any other part) and each such that $\alpha = 60^\circ$ (decided by the values of C and R for a given $\omega = 2\pi f$) enables the feedback voltage across the base and earthed (common) emitter to be 180° out of phase with that between the collector and the emitter, so that the voltage due to feedback which appears at the collector is 360° out of phase with that due to any signal voltage applied directly to the base of the bipolar transistor. This means that the feedback is positive. The frequency f of the oscillation provided by the circuit of Fig. 3.15 is given by

$$f = 1/(2\pi 6^{\frac{1}{2}} RC)$$

The quality of the sinusoidal waveform obtained can be examined by a cro connected as shown in Fig. 3.15.

3.9.2 *A crystal-controlled sinusoidal oscillator.* In Fig. 3.16, a quartz crystal (a 465kHz crystal is a good choice for an experiment) is connected between the collector and base terminals of the bipolar transistor whereas

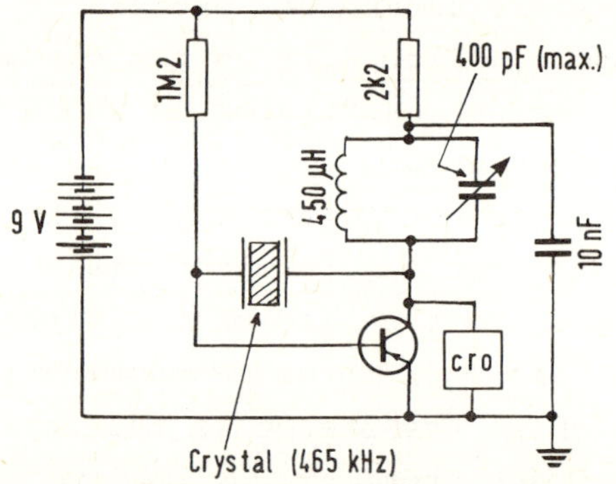

Fig. 3.16 A crystal-controlled sinusoidal oscillator based on a bipolar transistor (eg Type OC84)

an LC oscillatory circuit (L = 450 µH and C = 400 µF max) is connected in series with a load resistance of 2.2 kΩ in the collector circuit of this bipolar transistor. The sinusoidal waveform provided by this circuit can be examined by means of a cro connected as shown in Fig. 3.16. To measure the frequency f of this type of oscillator (or of the phase-shift oscillator described in Section 3.9.1) a digital frequency meter is a good choice.

3.9.3 *A Wien bridge oscillator.* An ac bridge introduced by Wien enables capacitance to be measured in terms of frequency and resistance. As the balance of this bridge circuit depends on the frequency, a sinusoidal oscillator can be based on it provided that the damping due to the resistance in the circuit can be overcome, usually by means of positive feedback, which requires a non-inverting amplifier, normally provided by two inverting stages coupled together in cascade. The amplifier must provide sufficient gain to compensate for the attenuation in the bridge network.

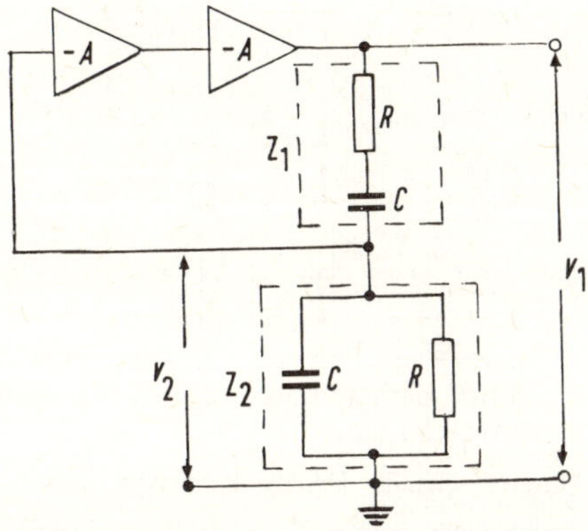

Fig. 3.17 Principle of a Wien Bridge oscillator

Consider the circuit of Fig. 3.17. The necessary non-inverting amplifier is provided by two amplifiers in cascade, each of which inverts to give a gain of $-A$. The input impedance to this amplifier must be large so that it has little effect on the impedances Z_1 and Z_2 of the Wien bridge network. The first amplifier cannot therefore simply be a bipolar transistor because its input impedance would be too low; instead, this first amplifier uses two bipolar transistors connected as a Darlington pair. The ratio of the voltages v_1/v_2 is given by

$$\begin{aligned} v_1/v_2 &= (Z_1 + Z_2)/Z_2 = 1 + Z_1/Z_2 \\ &= 1 + (1/R + j\omega C)/(R + 1/j\omega C) \\ &= 3 + j\,(\omega CR - 1/\omega CR) \end{aligned} \tag{3.11}$$

Provided that the ratio v_1/v_2 is positive, these voltages v_1 and v_2 are in phase when the imaginary part of this equation is zero; thus, $\omega CR = 1/\omega CR$, so that

$$f = \omega/2\pi = 1/2\pi CR \tag{3.12}$$

where f is the oscillator frequency.

The voltage attenuation produced by the network is given by the ratio of the amplitudes of the in-phase components, which is seen from equation (3.11) to be

$$v_1/v_2 = 3.$$

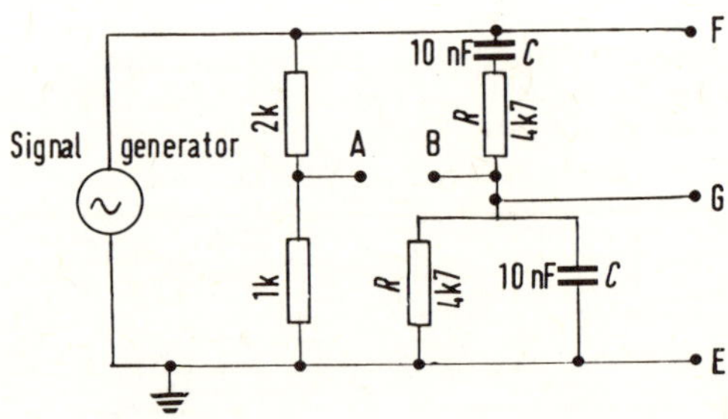

Fig. 3.18 The behaviour of a Wien bridge at various frequencies.

In an experiment on a Wien bridge with the circuit of Fig. 3.18, connect an ac valve voltmeter between the terminals A and B. Set a signal generator as shown to provide a sinusoidal output voltage of amplitude 5V peak-to-peak at a frequency of 200 Hz and record the voltage V_{AB} across AB. Increase in steps the frequency of this signal generator up to 20 kHz and record the voltage V_{AB} at each setting.

Plot a graph (using logarithmic scales) of V_{AB} against frequency. Repeat this procedure with the valve voltmeter connected to record V_{BE}, the voltage across B and E, and plot a graph of V_{BE} against frequency. Note that the maximum voltage across BE occurs at the same frequency as the minimum voltage across AB.

Connect the terminals FE to the *y* amplifier of a cro and the terminals GE to the *x* amplifier; observe the phase relationship between the voltages across FE and GE as the signal frequency is varied. When the bridge is balanced, the two voltages are in phase, the trace on the cro is a straight line and it should be checked that (in accordance with equation 3.11),

$$f = 1/2\pi CR$$

Set up the Wien bridge oscillator circuit of Fig. 3.19; on a cro display and record the output waveform. Measure the output frequency (or time period) with a digital frequency meter. Vary R_1 between 30 Ω and 10 kΩ (where, at any one time the two resistances labelled R_1 in Fig. 3.19 should be equal) and measure the frequency corresponding to each value of R_1. Plot a graph of time period T (= $1/f$) against R (=R_1 + 680) in order to verify equation (3.12). Note that with the capacitance C = 100 nF and the resistance R_1 variable up to a maximum of 10 kΩ, the frequency range covered is 150 Hz to 2kHz.

Repeat the experiment with the values of C set at 10 nF and at 1000 pF.

Note that, in this circuit (Fig. 3.19) the input stage of the amplifier is a Darlington pair (type 2N5306) which, as for all Darlington pairs, has a very high current gain (h_{FE} is typically > 10^4), and a very high input impedance (h_{ie} is typically 650 kΩ). Furthermore, the 330 Ω resistor in the emitter lead of the 2N5306 is not bypassed so that negative feedback will further increase the input impedance of the amplifier and reduce the voltage gain to approximately.

$$R_C/R_E = 3300/330 = 10$$

Another feature of this circuit (Fig. 3.19) is the negative temperature-coefficient thermistor R53 which provides negative feedback and amplitude control. This directly-heated bead-type thermistor is mounted in an evacuated glass tube and has a resistance of approximately 5 kΩ at 20°C. When the circuit begins to oscillate, the amplitude of the oscillations would increase unless limited in some manner. As the current increases through

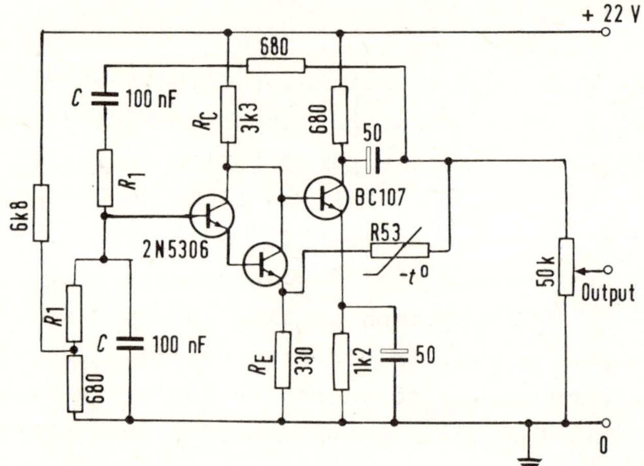

Fig. 3.19 A Wien bridge oscillator based on the use of bipolar transistors of which the first two (eg 2N5306) form a Darlington pair.

the thermistor R53, its resistance decreases, the feedback fraction is increased and so the amplifier gain is automatically reduced.

Set up the circuit of Fig. 3.19 with C = 10 nF and adjust the R_1 values until stable oscillations are obtained at a frequency of 10 kHz approximately. Measure the amplitude of the output with a valve voltmeter and its frequency with a digital frequency meter. Set up an amplifier circuit which is the same as that of the amplifier in Fig. 3.19 and provide it with an input signal voltage v_i from a signal generator set at the same frequency as that at which the Wien bridge oscillator was operating.

Increase the amplitude of the input signal voltage v_i until the output amplitude is equal to that measured with the oscillator in the stable state. Determine the voltage gain (v_o/v_i) of the amplifier and compare it with the value of 3 predicted earlier in this section.

3.10 *Multivibrators*

There are three basic classes of multivibrator circuits: the stable or free-running, the monostable and the bistable. Although it could be claimed that the term 'multivibrator' is a misnomer when applied to the monostable and bistable circuits which are really switch circuits rather than ones which generate continuously an output they are usually classified as types of multivibrator circuit.

3.10.1 *The astable (free-running) multivibrator.* This has two quasistable states and periodic transitions occur from one state to the other. Referring to Fig. 3.20 (in which T_1 and T_2 are bipolar transistors) one state is with the transistor T_1 fully conducting (saturated) and the other transistor T_2 non-conducting (cut-off) whereas the other state is the reverse of this (ie T_1 is at cut-off and T_2 is saturated). Because each transistor switches rapidly from a fully conducting state and back again, the output voltage waveform (observed by a cro connected across points A or B and earth in the circuit of Fig. 3.20) is approximately rectangular. Such a waveform can be shown to contain a large number of sinusoidal components covering a very wide range of frequencies - hence the name 'multivibrator'. If, in Fig. 3.20, $R_1 = R_4$, $R_2 = R_3 = R$ and $C_1 = C_2 = C$, the circuit is said to be symmetrical and each transistor remains in a conducting state for the same time. In this case, the 'mark-to-space' ratio (ie the ratio of the pulse duration to the time interval between pulses) is unity and the time period T of the oscillator is given approximately by

$T = 1.4RC$ s

or the frequency $f = 1/(1.4\ RC)$ Hz (3.12)

where R is in ohms and C in farads.

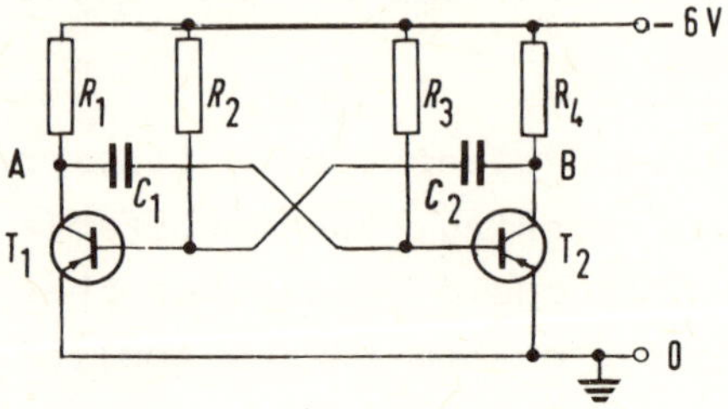

Fig. 3.20 A free-running multivibrator based on bipolar transistors.

In an experiment with the circuit of Fig. 3.20 make $R_1 = R_4 = 1$ kΩ, $R_2 = R_3 = 82$ kΩ and $C_1 = C_2 = 0.0047$ μF. Use p-n-p bipolar transistors of type ACY21 for both T_1 and T_2 and a 6 V battery for the power supply. Observe and record the voltage waveforms $(V_{ce})_{T_1}$ and $(V_{be})_{T_1}$ with a double-beam cro. With $R_1 = R_4 = 1$ kΩ, vary R_2 and R_3 or C_1 and C_2 (keeping the circuit symmetrical) and measure with a cro the frequency for each set of values. Plot a graph to verify equation (3.12).

Investigate the effect of a change in temperature on the frequency of this multivibrator by warming the bipolar transistors slightly by playing on to them hot air from a hair dryer.

3.10.2 *Effect of varying the applied voltage on the period of a free-running multivibrator.* Keep the 6 V battery supply of Fig. 3.20 but arrange a variable voltage supply to the ends of R_1 and R_3 as shown in Fig. 3.21. With $R_1 = R_4 = 1$ kΩ, $R_2 = R_3 = 82$ kΩ, $C_1 = C_2 = 0.0047$ μF, investigate the effect on the frequency (the reciprocal of the period), of varying the applied voltage V over the range 3 - 25 V whilst keeping the voltage V_{CC} (see Fig. 3.21) at -6V. Calculate a value for the voltage sensitivity df/dV.

With this circuit in which the applied voltage V is varied, the time period T of the oscillations is given by

$$T = 2RC \log_e (1 + V_{CC}/V)$$

Therefore

$$f = 1/T = 1/ | 2RC \log_e (1 + V_{CC}/V) |$$

If V_{CC}/V is small,

$$f = V/2RCV_{CC}$$

so that

$$f \propto V.$$

Plot a graph of frequency against voltage and examine the range over which the approximation $f \propto V$ is valid. Note that an elementary type of voltage-to-frequency converter is made available.

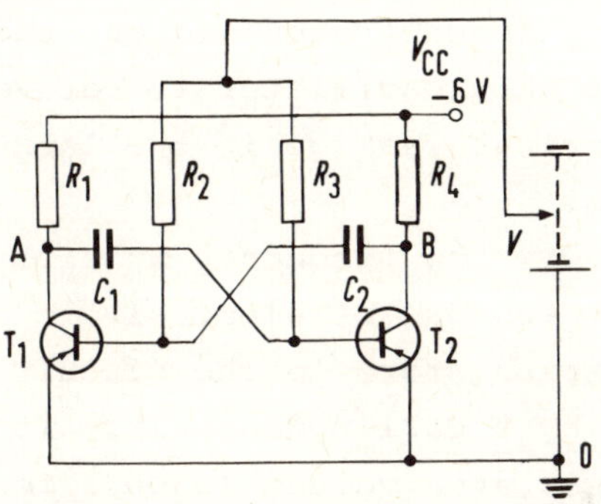

Fig. 3.21 Effect of varying the applied voltage on the period of a free-running multivibrator (T_1 and T_2 are, eg, each ACY21)

A very low frequency free-running multivibrator. Set up the circuit of Fig. 3.22 in which large value (100 μF) electrolytic capacitors are used for C_1 and C_2 to enable the time period (with $R_1 = R_2 = 10$ kΩ) to be made about 2 s , so that the switching action can be readily observed by means of two small filament lamps L_1 and L_2 or by connecting a moving-coil voltmeter across either of the bipolar transistors to record the variation of V_{ce} Note that the positive plate of each electrolytic capacitor in

Fig. 3.22 is connected to the base of the appropriate transistor.

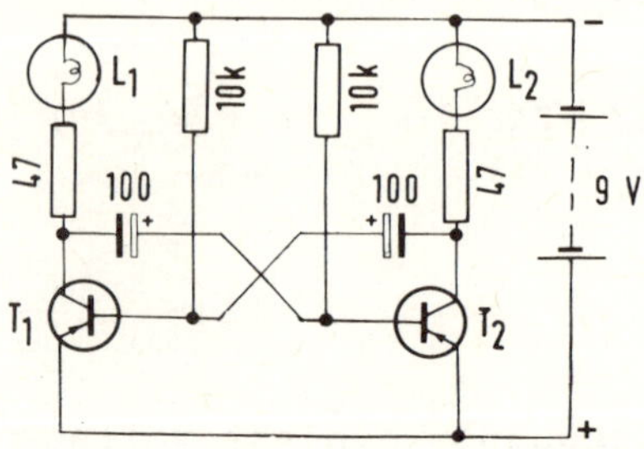

Fig. 3.22 A very low frequency multivibrator based on bipolar transistors. L_1 and L_2 are both 6 V, 0.06 A light bulbs

Although this experiment enables the switching action to be graphically demonstrated without the use of a cro, the frequency of this oscillator is not very constant because of electrical leakage in the large value electrolytic capacitors.

To enable a very low frequency multivibrator to be constructed without the use of electrolytic capacitors,each of the bipolar transistors in the circuit of Fig. 3.22 may be replaced by a Darlington pair. It is then possible to use large resistance values (each a few MΩ) and obtain a low frequency (f < 1Hz) with capacitors of values up to 1 μF.

3.10.3 *A gated free-running multivibrator.* In the circuit of Fig. 3.20 a third bipolar transistor T_3 is placed in the emitter lead of either T_1 or T_2, and a separate voltage supply is used to hold T_3 either ON or OFF. This gives a 'gated multivibrator' of which the oscillations can be started or stopped in a convenient way.

3.10.4 *The monostable multivibrator.* This is a circuit which has one stable and one semi-stable state. When the supply voltage is connected, the circuit adopts the stable state and remains in it until an externally applied pulse (the switching signal) switches the circuit to the semi-stable state. The time interval during which the circuit remains in the semi-stable state is determined by the time constant of an *RC* network. At the end of this time interval, the circuit reverts to its stable state and remains there until a second switching signal causes it to make another temporary transition.

This circuit is sometimes called a 'one-shot multivibrator'. The trigger action (by which it is implied that the leading edge of the switching signal pulse is used only to initiate the transition, and subsequently the behaviour is decided by the values of the components in the monostable circuit) of this circuit arrangement is of special interest because the switching signal can have a variety of forms provided that its amplitude is sufficient to initiate the transition.

The output pulse waveform, on the other hand, is always rectangular with an amplitude which is independent of the amplitude or duration of the input signal. Herein lies the value of the monostable multivibrator circuit: it provides a rapidly acting switch with a specifically shaped output pulse, whereas the input pulse can be of irregular shape provided that its amplitude is big enough. For this reason, this type of circuit is often used as a discriminator which ignores spurious low amplitude noise pulses whereas it transmits genuine pulses of amplitude exceeding some threshold value. The monostable circuit can also be used to undertake the following operations.

(i) Introduce a time delay (often of the order of milliseconds) between the receipt of the initiating switching signal and the transition back to the stable state.

(ii) Shape pulses

(iii) Provide a rectangular waveform-generator of which the pulse duration can be varied.

(iv) Open and close a switch for a chosen time interval.

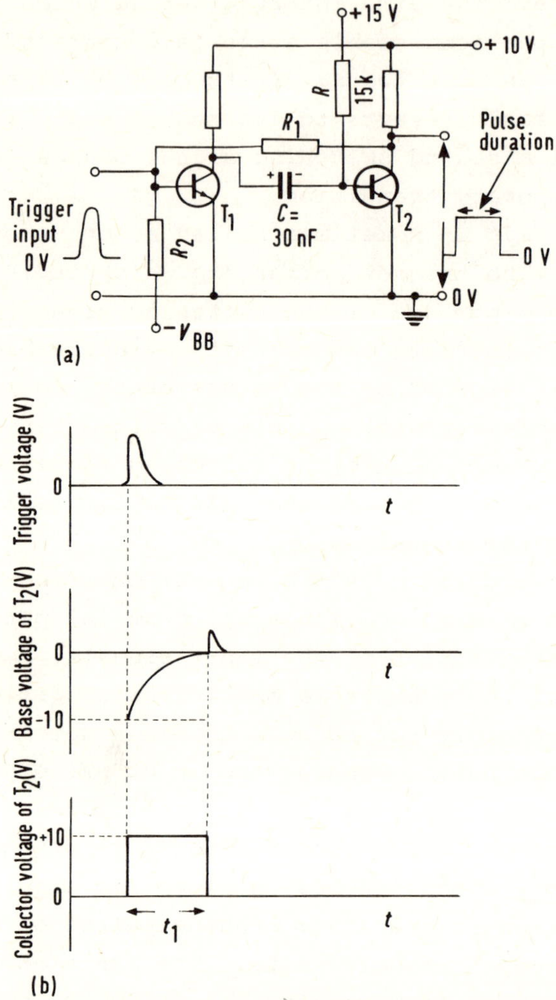

Fig. 3.23 A basic monostable multivibrator utilizing bipolar transistors (a) circuit diagram; (b) waveforms illustrating operation.

In a basic simple monostable circuit which makes use of two bipolar transistors (Fig. 3.23) the stable state exists with transistor T_2 conducting and transistor T_1 at cut-off. The base current of T_2 is provided from the 15 V supply via the 15 kΩ resistor *R*. When transistor T_2 is at saturation, its collector may be assumed to be at earth potential and the capacitor C is charged (with the voltage supply values shown in Fig. 3.23a) to 10 V with the polarity indicated. The reverse bias (at the cut-off value) to transistor T_1 is provided by the voltage $-V_{BB}$ in conjunction with the potential divider resistors R_1 and R_2. When a positive pulse is applied to the base of T_1, this transistor is switched ON whereas T_2 is turned OFF. Positive feedback (because the output of T_2 is connected to the input of T_1) creates regenerative switching thus ensuring that the transition is rapid. Transistor T_1 conducts at saturation so it connects to earth the positive plate of the capacitor C which therefore discharges through R. Transistor T_2 remains at cut-off until the voltage on C reaches zero, when the transition back to the stable state will occur. The time interval for which T_2 is held at cut-off determines the duration of the output pulse.

The typical waveform for a monostable circuit (Fig. 3.23b) illustrates the duration (t_1) of the output pulse (often called the pulse width) which may be calculated from the capacitor discharge curve (Fig. 3.24) which has

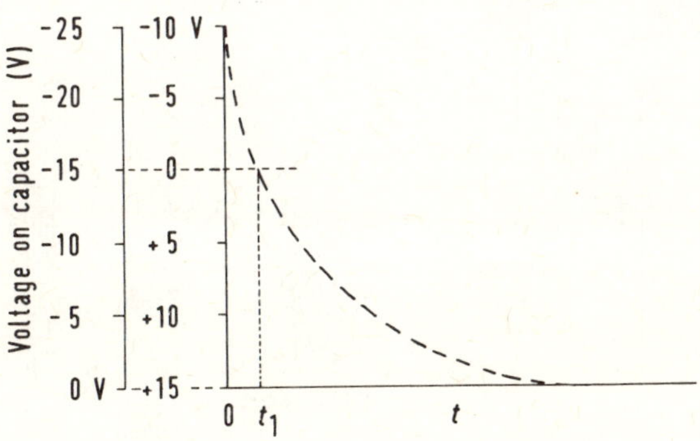

Fig. 3.24 Capacitor discharge curve and calculation of the pulse duration t_1.

a false zero marked on the voltage axis. Hence, t_1 is given by

$$-15 = -25 \exp(-t_1/RC)$$

therefore

$$t_1 = 15\times10^3 \times 30\times10^{-9}\log_e 1.67$$
$$= 231\ \mu s.$$

3.10.5 *A monostable multivibrator utilizing bipolar transistors for experimental work.* The circuit used (Fig. 3.25) adopts the usual practice whereby the ON transistor is turned OFF, rather than adopting the procedure indicated in relation to Fig. 3.23a. The circuit of Fig. 3.25 is constructed making use of bipolar n-p-n silicon transistors (type 2N 3094 is suitable).

Before any external pulse is applied, measure the steady voltages at a number of selected points to establish that T_2 is cut off. Connect a signal generator providing a square waveform at a frequency of 400 Hz between the trigger input terminal and the earth line. The negative half-cycle of this input can be used to initiate the transition to the semi-stable state. Display and record the output waveforms on a cro. Record the minimum amplitude of the pulse required to cause a transition. Observe the

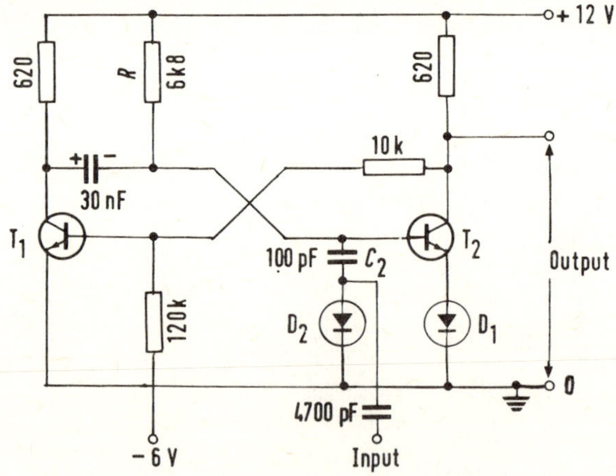

Fig. 3.25 A monostable multivibrator utilizing bipolar transistors for experimental work.

effect on the output of varying the amplitude and the frequency of the trigger signal.

This monostable circuit provides a rectangular output pulse of duration (or width) t given by

$$t = 0.69\ RC\ \mathrm{s}$$

if R is in ohms and C is in farads. Measure the output pulse width for a range of values of R and C. Plot a graph of the observed pulse width t_{obs} against the calculated pulse width t_{calc}.

3.10.6 *The bistable multivibrator.* A bistable circuit is one which can exist indefinitely in either of two stable states and where an abrupt transition from one state to the other can be induced by means of some external excitation usually in the form of an input pulse. The bistable circuit is used extensively for counting and for storing binary information. Other terms for the basic unit include 'scale-of-two', 'flip-flop' and 'binary'.

The basic bistable multivibrator circuit employs positive (regenerative) feedback during the interval required for the switching to take place. This regenerative action will not occur until initiated by an externally applied trigger voltage. The basic circuit (Fig. 3.26a) is drawn as a two-stage amplifier based on bipolar transistors T_1 and T_2 with direct coupling between the stages. In Fig. 3.26b the same circuit is re-drawn to emphasise the symmetry of the arrangement.

The possibility that equal currents will flow in transistors T_1 and T_2 in some initial state is very remote. Thus, consider a small voltage fluctuation at the base of transistor T_1. This fluctuation will be amplified and fed in an inverted state to the base of T_2, where it is further amplified by T_2 and inverted to be fed back to the base of T_1 where the original voltage fluctuation is augmented. This regenerative feedback will ensure that one of the transistors is driven into saturation while

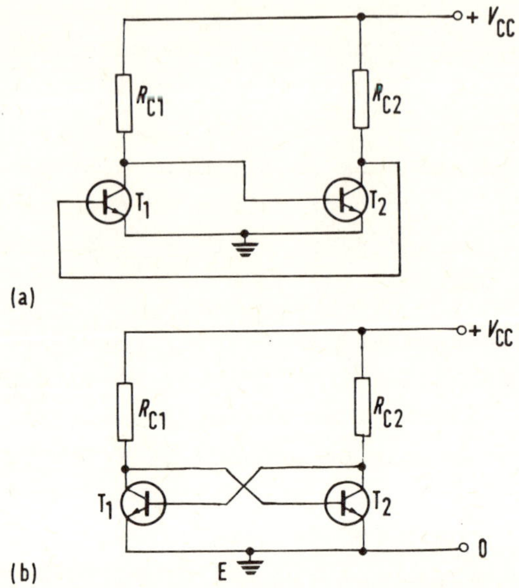

Fig. 3.26 The bistable multivibrator: (a) a two-stage amplifier with directly-coupled positive feedback; (b) the same amplifier circuit redrawn to emphasise the symmetry of the arrangement.

the other is cut off. The circuit will remain in this stable state until an input pulse initiates the transition.

The capacitors in the developed circuit (Fig. 3.27) are termed 'commutating' or 'speed-up' capacitors. They serve to reduce the transition time, which is the time interval during which conduction is transferred from one transistor to the other. Increasing the capacitance of these capacitors will reduce the transition time. However, a certain minimum time (the resolving time) must elapse after one switching sequence before a succeeding trigger pulse will reliably induce the reverse state. This resolving time is the sum of the transition time and the settling time of the circuit. Increasing the capacitance increases the settling time: a compromise is therefore required in the choice of the capacitance C for optimum behaviour.

The resolving time of the bistable multivibrator is the

reciprocal of the maximum frequency at which the circuit will respond.

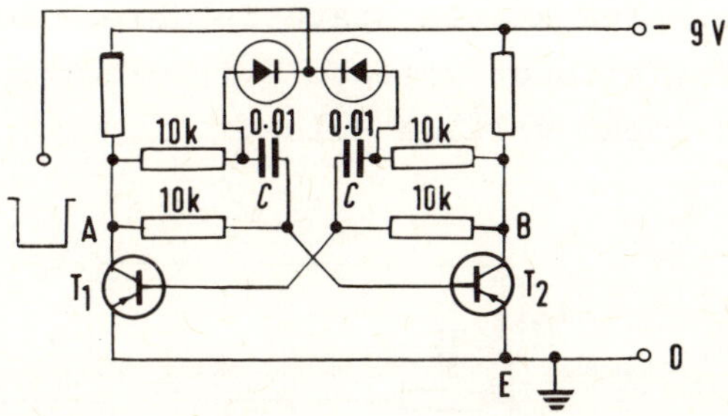

Fig. 3.27 A bistable multivibrator for experimental work. (T_1 and T_2 can be, eg, ACY21).

3.10.7 *A bistable multivibrator based on bipolar transistors.* With the circuit of Fig. 3.27 a pulse generator is used to provide negative-going variable-height pulses of which the heights are measured by a voltage-calibrated cro. The minimum pulse height is found which enables the circuit to be switched from one stable state to the other.

With this same circuit find out whether or not the shape of the triggering pulse is important for reliable switching. To do this, make use of a sinusoidal voltage source and any other pulse source available (eg from a unijunction relaxation oscillator, Section 5.2).

Use filament lamps as loads and cause the circuit to switch at very low frequencies.

Observe and record the waveforms of the voltages with respect to earth at the points A and B.

Measure the maximum frequency at which the multivibrator will switch reliably and hence calculate the resolving time of the circuit.

Operate this bistable circuit at very low frequency and use suitable light-emitting diodes in place of the filament lamps used previously.

3.11 *The Schmitt trigger circuit*

This circuit - named after the inventor of the vacuum-tube version - is a special form of the emitter-coupled bistable multivibrator. Its stable state is determined by the amplitude of the input voltage - in practice usually the height of the input voltage pulse.

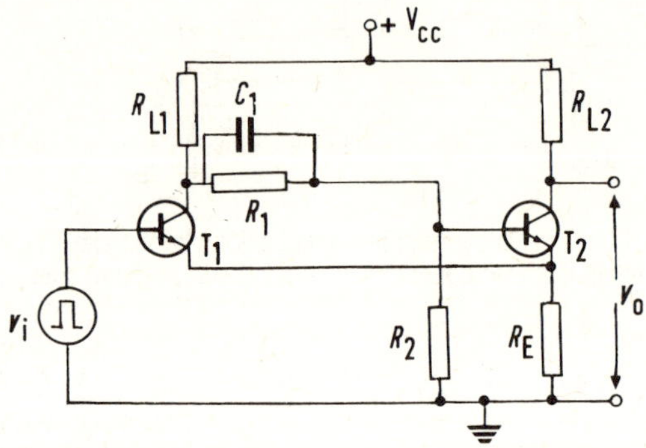

Fig. 3.28 The Schmitt trigger circuit based on bipolar transistors.

In the Schmitt trigger-circuit, based on bipolar transistors T_1 and T_2 (Fig. 3.28), when the input voltage v_i applied is zero, transistor T_1 is at cut-off and transistor T_2 conducts at saturation. Before T_1 can be made to conduct, the input voltage v_i must exceed the voltage drop of V_{E2} across R_E. The amplitudes of the input voltage needed to cause T_1 to conduct is called the upper trigger potential (utp). As T_1 starts to conduct, the forward bias on T_2 is reduced and a very rapid transition occurs so that T_1 conducts at saturation and T_2 is at cut-off. This state will continue whilst the input voltage is greater than the utp. If the input voltage v_i is then reduced, the circuit will not switch back again to its initial state (with T_1 at cut-off and T_2 at saturation) until the input voltage is reduced to the lower trigger potential (ltp) which is less than the utp. If the input signal is sinusoidal, the switching produced is shown in Fig. 3.29 in which is also shown the switching produced by an input pulse with some damped oscillation on its flat top. This diagram also shows the shape of the pulse of the output voltage v_o. The circuit is said to exhibit hysteresis of magnitude V_H = UTP - LTP.

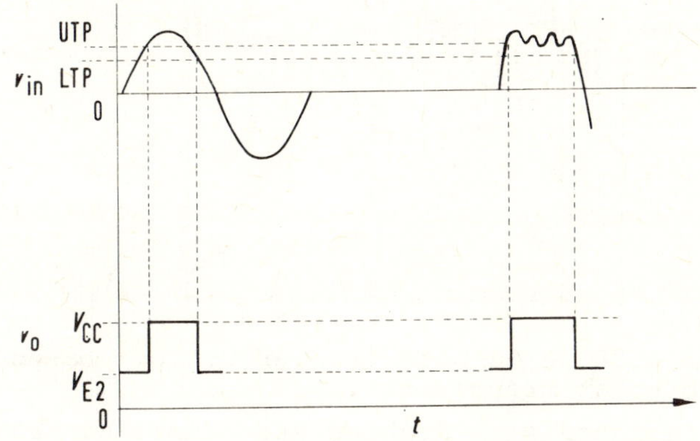

Fig. 3.29 The output from a Schmitt trigger circuit for two different input waveforms.

An important application of the Schmitt circuit is to use it as an amplitude comparator to mark the instant at which some arbitrary waveform attains a particular reference level. In a second application the circuit acts as a pulse shaper or a squaring circuit. Provided that the excursions of the input signal are large enough to exceed the limits of V_H the output voltage v_o is a square wave of the same period as the input and amplitude independant of the amplitude of the input.

In an experiment, set up the circuit of Fig. 3.30 in which a potential divider network is provided at the input to the transistor T_1. By means of a digital voltmeter measure the steady input voltage between the base of T_1 and earth which is required to make T_1 conduct at saturation, thus measuring the utp of the circuit, and also measure the voltage at which the circuit reverts to T_2 conducting at saturation, ie measure the ltp. Measure of the voltage (by means of a separate digital voltmeter) across

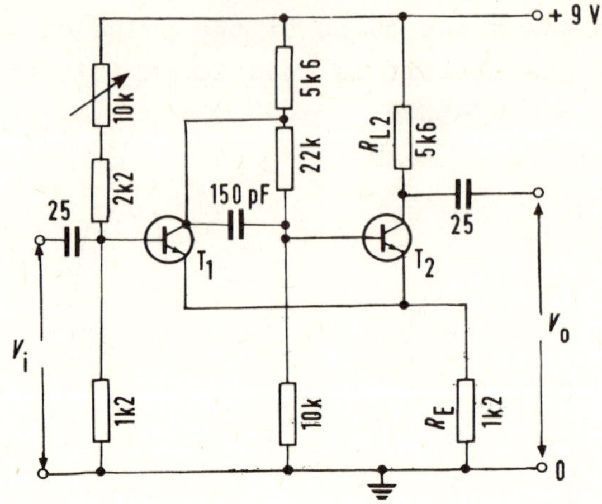

Fig. 3.30 A Schmitt trigger circuit based on bipolar transistors.

the load resistor R_{L2} enables the points of transition to be determined.

With the input potentiometer set so that the transistor T_1 is biased at approximately 1 V below the ltp, examine the output waveforms for a number of input signals of different kinds. With a cro, display and record the input and output waveforms simultaneously.

Ensuring that R_E exceeds 470 Ω, vary its value and investigate the effect of this change on the utp.

The potential divider network governing the potential at the base of T_1 can be used to arrange for this circuit to be switched by pulses of a specific amplitude. Find the minimum pulse height required to switch this circuit.

Determine the maximum frequency at which this circuit will behave satisfactorily

3.12 *Sweep generator: utilizing the bootstrap principle*

A voltage waveform displays voltage as a function of time. Sweep generators are based on circuits which produce repetitive waveforms which, for the major portion of any cycle, show a linear variation of voltage with time.

Synonymous with the term 'sweep generator' are the terms 'ramp generator' and 'time-base generator'. An important method of producing a sweep generator is by making use of the bootstrap principle. This is illustrated by the circuit of Fig. 3.31.

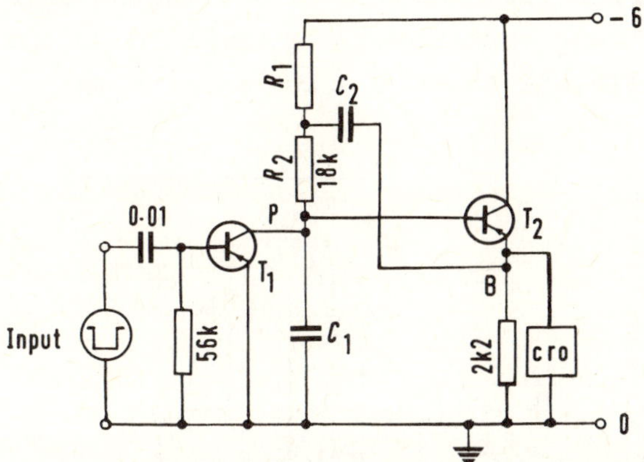

Fig. 3.31 Linear ramp circuit based on the use of the bootstrap principle with bipolar transistors.

Capacitor C_1 would charge exponentially through R_1 and R_2 from the 6 V supply if the emitter-follower connected transistor T_1 circuit were absent. The pd so developed across C_1 causes the potential at the junction between R_1 and R_2 to become more negative. This change of voltage is followed at point B because B is connected to the junction. Because the potential at B is a constant fraction of that at P, the pd across R_2, and the current through it, are both constant.

In an experiment, set up the circuit of Fig. 3.31 with R_1 = 18 kΩ and C_2 = 1 μF. Use a pulse generator (which provides negative-going square pulses of controllable height and frequency) to apply negative-going pulses to the base of the transistor T_1. Display the output voltage waveform on one beam of a cro (connected as shown in Fig. 3.31) whilst the input pulse waveform is displayed by means of the other beam. Note the effect of varying the number of input pulses per second.

Use a number of values of the capacitance C_2 over the range from 0.02 - 100 μF and for each record the waveform of the output voltage with a cro. Repeat using the circuit of Fig. 3.32 based on a Darlington pair in emitter-follower connection. Attempt to deduce the optimum conditions for the most linear ramp.

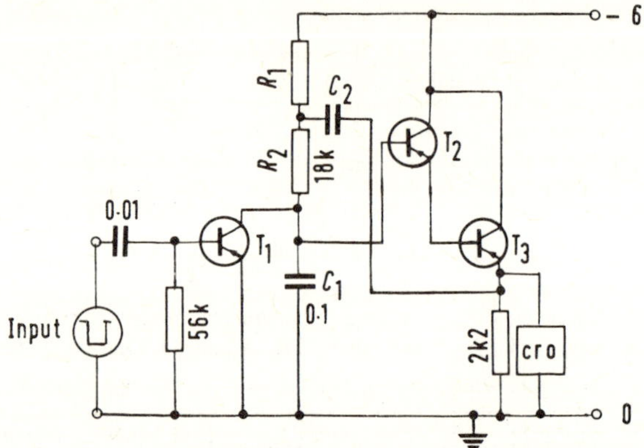

Fig. 3.32 Linear ramp generator based on the bootstrap principle with a Darlington pair (T_1 and T_2) in the emitter-follower connection.

3.13 *An optically-coupled isolator*

The optically-coupled isolator used is a TIL112 manufactured by Texas Instruments; it is mounted within a six-pin dual in-line plastic package. A spot on the top surface is adjacent to pin number 1. The package contains a gallium arsenide infra-red light-emitting diode (LED) coupled by the radiation it emits with a silicon n-p-n phototransistor (Fig. 3.33). The coupling between the LED and the phototransistor is solely by the infra-red radiation emitted by the LED. This optical coupling enables information (signals) to pass from one circuit to the other even though the circuits are electrically isolated. The quality of an optical coupler is assessed in terms of the following features:

(i) Isolation: typically either circuit can be operated at a potential of 1.5 kV with respect to the other.

(ii) Current transfer ratio: the ratio of the collector current (from the phototransistor) to the input current to the LED should be high as possible (a value of 0.2 is typical).

(iii) Switching speed: a rise-time of 2 μs is typical.

(iv) Isolation resistance: generally exceeds 10^{11} Ω.

(v) Intercircuit (input-to-output) capacitance: typically of the order of 1 pF.

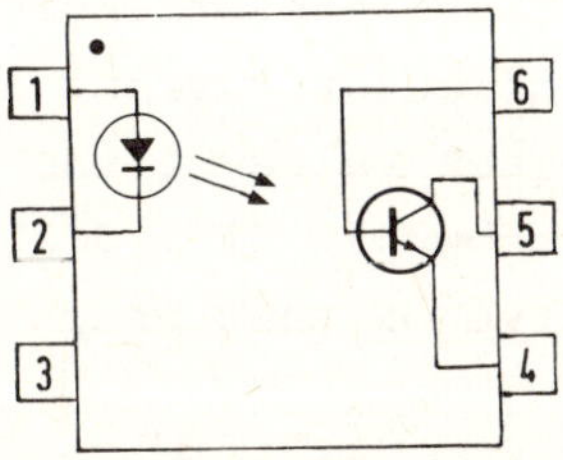

Fig. 3.33 Pin configuration and internal structure of a typical optically-coupled isolator (TIL112), viewed from above.

3.13.1 *Output characteristics of the phototransistor.* Set up the circuit shown in Fig. 3.34. By varying the value of the variable resistance R_1 in series with the light-emitting diode (LED), arrange for the current I_F through the diode to be 5 mA. Increase the collector-emitter voltage V_{CE} of the phototransistor in steps between 0 and 25 V; at each setting of V_{CE} record the collector current I_C. Repeat with the diode current I_F set at 10 mA, 15 mA and 20 mA.

Note that the maximum continuous power dissipation in the phototransistor should not exceed 150 mW; ie the diode current I_F should not exceed 15 mA (see Fig. 3.35).

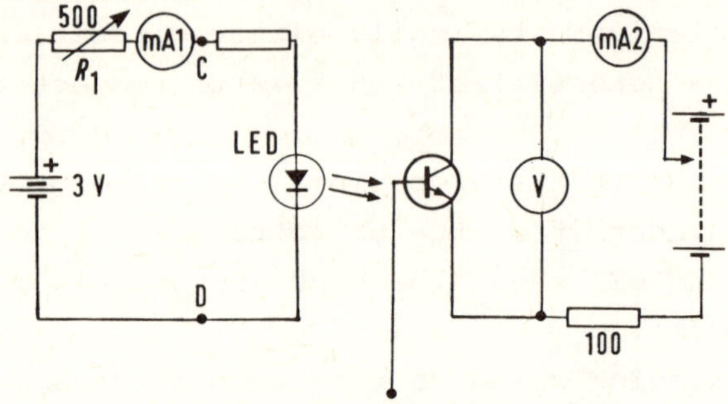

Fig. 3.34 Determination of the output characteristics of the phototransistor of the optically-coupled isolator TIL112. mAI is a milliammeter with fsd 25 mA, measures LED current I_F; mA2 is a milliammeter with fsd 10 mA, measures phototransistor collector current; V is a voltmeter with fsd 25 V, measures V_{CE} of the phototransistor.

Plot a graph of collector current against collector-emitter voltage for each value of the diode current I_F. Typical results are shown in Fig. 3.35. Draw the maximum power dissipation curve on this graph.

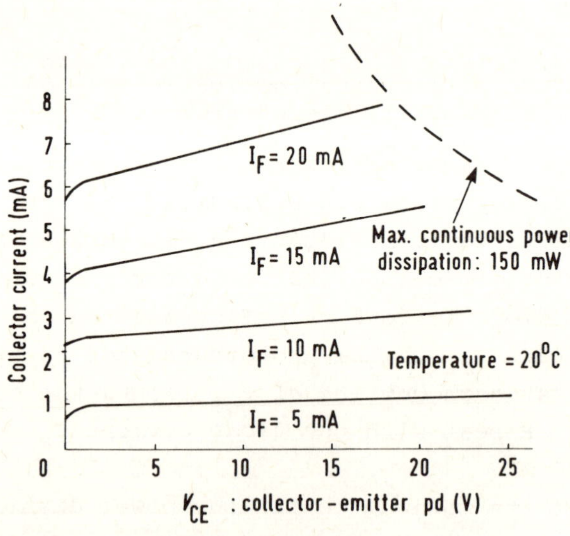

Fig. 3.35 Collector current I_C against collector-emitter voltage V_{CE} for the phototransistor of the TIL112.

3.13.2 *Current transfer ratio.* Using the same circuit as in the previous experiment (Fig. 3.34) set V_{CE} at 5 V. Increase the forward current I_F through the LED in increments of 2 mA from 0 to 20 mA and at each current setting record the collector current I_C. Plot a graph of I_C against I_F.

With V_{CE} still maintained at 5 V, it is profitable to examine experimentally in more detail the relationship between I_C and I_F over the range 0 - 5 mA for the diode current I_F.

3.13.3 *Photodiode operation.* It is possible to utilise the collector-base junction of the n-p-n transistor as a photodiode. With a reverse bias across the photodiode and no radiation incident on the junction, the 'dark current' is negligible. Current will flow when electromagnetic radiation of appropriate wavelength (ie photons of the appropriate energy) falls on the junction.

Set up the circuit shown in Fig. 3.36. With a reverse pd or 10 V across the collector-base junction and $I_F = 0$, attempt to measure the 'dark current'. Increase I_F in increments of 2 mA between 0 and 20 mA; at each setting record the photodiode current. Plot a graph of photodiode current against forward LED current.

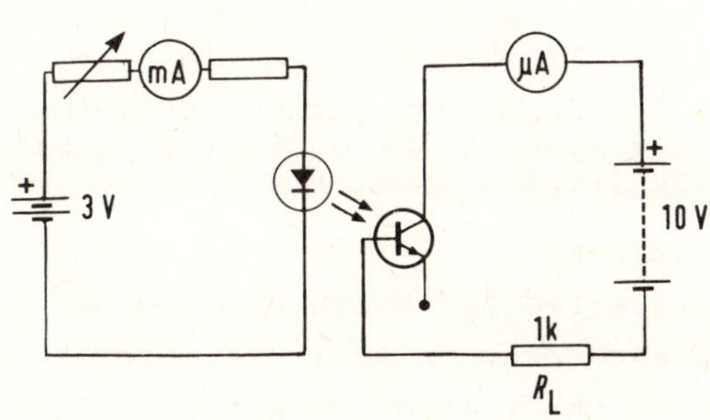

Fig. 3.36 Operation of the collector-base junction of the optically-coupled isolator (TIL112) as a photodiode.

3.13.4 *Switching characteristic.* Again making use of the circuit shown in Fig. 3.34 set V_{CE} = 10 V. Adjust the current I_O = 2 mA. To achieve this, measure the pd across the points CD, where this pd will be 1.7 V approximately for I_C = 2 mA. Disconnect the components to the left-hand side of CD and across CD connect a square-wave generator with an output resistance of approximately 75 Ω to provide a signal of amplitude 1.7 V. Remove the meters from the phototransistor circuit and display, on a double-beam cro, the input voltage waveform and the output voltage waveform across the 100 Ω resistor *R* in the emitter lead of the phototransistor. Typical waveforms are shown in Fig. 3.37. Record these waveforms and measure the rise-time t_r and the fall-time t_f, each of which should be of the order of 2 μs.

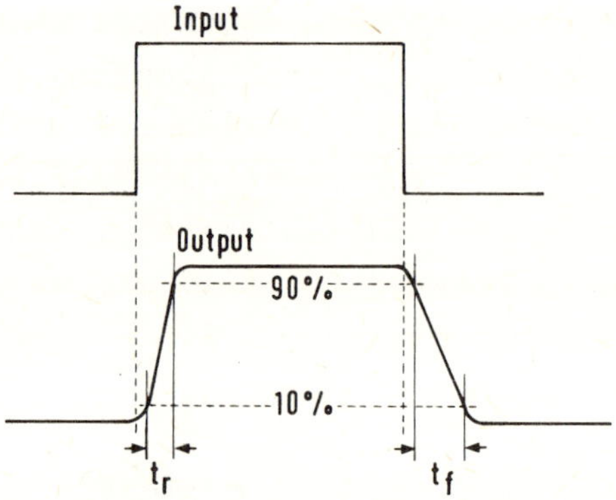

Fig. 3.37 Input and output waveforms in the switching mode of the optically-coupled isolator (TIL112).

3.13.5. *Questions*

(a) Is infra-red radiation emitted by the LED when it is forward-biased or when it is reverse-biased?

(b) What is the voltage drop across the forward-biased LED when a current of 5 mA is flowing?

(c) Outline some of the advantages of a solid-state optically-coupled isolator over a filament lamp and photocell arrangement.

3.14 *A typical application of an optically-coupled isolator*

The advantage of this device is that it can be used to observe a signal in a circuit without electrical contact between the source of this signal and the output from the phototransistor of the isolator. For example, to examine the switching action waveforms of a free-running multivibrator based on two bipolar transistors (as in Section 3.11.1) the LED of the isolator (the TIL112, for example) is connected (with its 47 Ω resistor in series) in series with the 1 kΩ resistor in the collector circuit of the bipolar transistor T_2 to give the circuit of Fig. 3.38. The switching action of this multivibrator may then be observed across the 100 Ω resistor in the emitter lead of the phototransistor of the isolator. The steady voltage level at the point from which the signal originates has no effect (because of the electrical isolation) on the voltage level at the output.

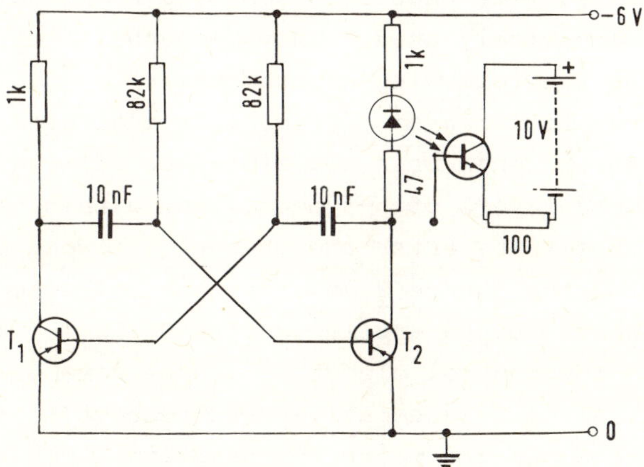

Fig. 3.38 An optically-coupled isolator in a multivibrator circuit.

CHAPTER FOUR

Field effect transistors: characteristics and simple associated circuits

4.1 *Field-effect transistors (FETs or fets)*

The structure and the circuit symbols of the junction-gate field effect transistors (fets or jgfets) are shown in Fig. 4.1.

Unlike the bipolar transistor in which both types of current carrier (electrons and holes) are concerned, the fet involves only one type of current carrier (electrons <u>or</u> holes) and is thus a <u>unipolar</u> transistor. Electrons are the carriers in an n-channel fet and holes in a p-channel fet. As the mobility of electrons in silicon is approximately twice that of holes, the n-channel type is the much more widely used. Only n-channel fets will be examined in the following experiments.

The schematic diagram of Fig. 4.1a indicates the ohmic contacts at the ends of a rod of n-type silicon, with heavily doped layers of p-type silicon within the sides of the rod. Electrons enter the channel (between the p-type layers) from the source S and they are collected at the drain D, where the pd applied across DS is V_{DS} and is positive. The control electrode is the gate G; normally the potential on G is negative with respect to that on the same source S, ie the pd V_{GS} is negative. For a given value of V_{DS}, as the voltage on the gate with respect to the source (ie V_{GS}) is increased as a negative magnitude the depletion layer at the reverse-biased p-n junction extends into the channel, reducing its effective width and therefore increasing the resistance of the channel. Hence, the channel current - normally called the drain current I_D - is a maximum when $V_{GS} = 0$; V_{GS} is rarely made positive.

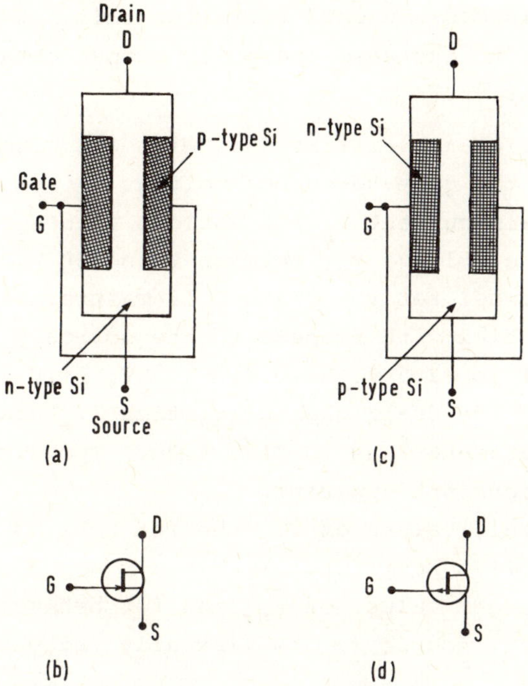

Fig. 4.1 Field effect transistors of the junction gate type: (a) structure of an n-channel fet; (b) circuit symbol for an n-channel fet; (c) structure of a p-channel fet; (d) circuit symbol for a p-channel fet.

The fet depends for its action on the control of the depletion region size. Used as a simple amplifier, the input signal voltage is normally applied in series with a steady bias voltage between the gate and the source. Since the steady bias is of such a polarity and magnitude that the junction is always reverse biased (ie the potential on G is negative with respect to that on S) the input resistance is very high. So the fet is often said to be a voltage-controlled device because negligible current is demanded from the input signal.

In general, a fet is a low power device used in the first stage of an amplifier to obtain low noise and virtual electrical isolation of the signal source. Often bipolar transistors are used to follow this fet input stage in order to increase the power output obtainable from the amplifier.

4.1.1 *The drain characteristics.* With the circuit of Fig. 4.2a set the gate-to-source voltage V_{GS} at -3 V and record the drain current I_D for various values of the drain-to-source voltage V_{DS} between 0 and 20 V. Note that with the n-channel fet the drain D is maintained at a positive potential with respect to the source S. Repeat with V_{GS} equal to -2.5 V, -2.0 V, -1.5 V, -1.0 V, -0.5 V and 0 V. Plot the drain characteristics (I_D against V_D with V_{GS} as parameter) as in Fig. 4.2b. Two distinct operating regions are apparent.

(a) For small values of V_{DS} the fet behaves like a voltage-controlled resistor.

(b) For larger values of V_{DS} the fet behaves like a constant current source (I_D is virtually independent of V_{DS}).

The fet is operated in the constant current region when it is used as an amplifier or as a switch. Some fets are specifically designed for use as voltage controlled resistors, and are sometimes referred to as field effect resistors, fers.

4.1.2 *Effect of temperature on the drain current* I_D. With $V_{GS} = 0$ and $V_{DS} = 20$ V, the drain current I_D is a maximum and the temperature of the fet will rise with the power dissipated in it. After operating the fet with its maximum drain current for several minutes, cool the metal can of the transistor with a piece of ice or a cold moist cloth, when the drain current I_D will be seen to increase

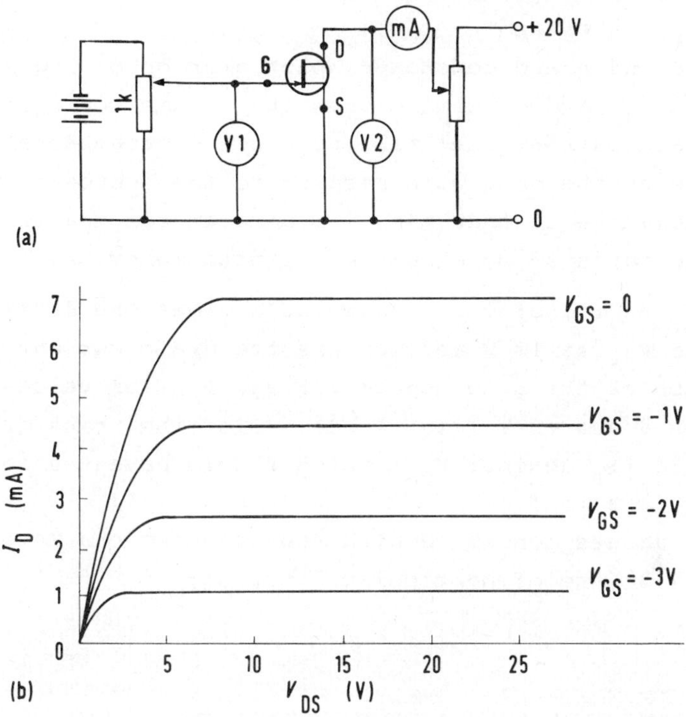

Fig. 4.2 (a) Circuit to obtain the characteristics of an n-channel fet. V1 is a high resistance voltmeter (preferably a dvm) with fsd of 5 V, which measures V_{GS} ; V_2 is a high resistance voltmeter (preferably a dvm) with fsd of 20 V, it measures V_{DS}; mA is a milliammeter with fsd of 10 mA, which measures I_D. (b) Typical drain characteristics for an n-channel fet.

as the temperature is decreased. Recalling that there is only one current carrier present and that the fet is made of silicon, what causes the drain current to rise as the temperature falls?

4.1.3 *Use as a voltage-controlled resistance.* Set V_{DS} = 2 V (Fig. 4.2a) and vary V_{GS} between 0 and -5V; record the drain current I_D for each setting of V_{GS}. Calculate the channel resistance $R_{DS} = \dfrac{dV_{DS}}{dI_D}$ for each value of V_{GS}.

Assuming the relationship

$$R_{DS} = C \exp(\eta V_{GS})$$

where C and η are constants, plot a graph of log R_{DS} against V_{GS} and determine from it values for C and η. The drain current will fall to a few microamperes as the voltage on the gate with respect to the source is made more negative so that the milliammeter for measuring I_D must be replaced by a suitable microammeter.

4.1.4 *The transfer characteristic.* Set the drain-source voltage V_{DS} at 15 V and measure the drain current I_D as a function of the gate-source voltage V_{GS} for values of V_{GS} between 0 and -5 V (Fig. 4.2a). Plot the transfer characteristic (I_D against V_{GS}) which should have the form shown in Fig. 4.3.

Two values concerned with the transfer characteristic (Fig. 4.3) are of particular interest;

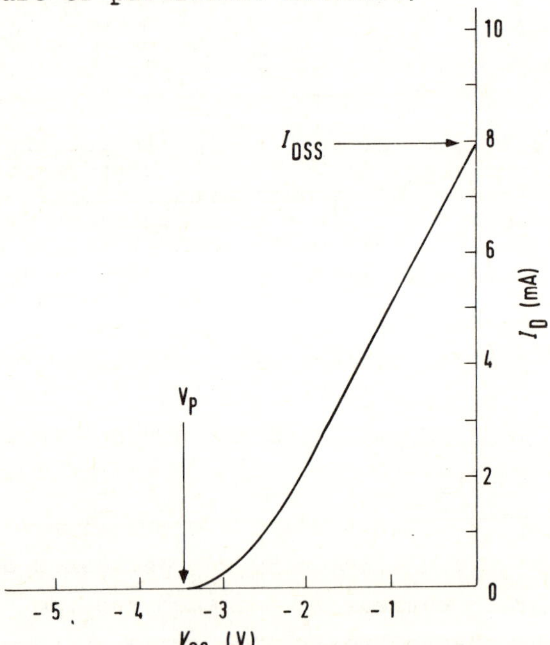

Fig. 4.3 A typical transfer characteristic for an n-channel fet.

(a) I_{DSS} : the drain current for zero bias ($V_{GS} = 0$),

(b) V_P : the gate-source voltage for zero drain current, sometimes called the 'pinch-off voltage'. The value of V_P is determined by extrapolating the curve. Manufacturers usually quote V_P for a specific value of I_D (eg for I_D = 1 nA).

Having determined the values of I_{DSS} and V_P, any other point (I_D, V_{GS}) on the transfer characteristic is given theoretically by the equation

$$I_D = I_{DSS}\,(1 - V_{GS}/V_P)^2 \tag{4.1}$$

This equation (4.1) suggests that the transfer characteristic is part of a parabola with its vertex at V_P. From the values of I_{DSS} and V_P found experimentally, calculate, by means of Equation (4.1), values of I_D corresponding to various values of V_{GS}. Tabulate these calculated values and plot from them the theoretical curve on the same axes as the experimental transfer characteristic.

On the transfer characteristic determined by experiment choose an operating point near the centre of the linear portion. Calculate the gradient $(\partial I_D/\partial V_{GS})_{V_{DS}}$ - the mutual conductance g_m of the fet - at this point.

An expression for g_m can be obtained by differentiation of Equation (4.1).

$$(\partial I_D/\partial V_{GS})_{V_{DS}} = g_m = -(2/V_P)\, I_{DSS}\, (1 - V_{GS}/V_P).$$

If V_{GS} is zero, g_m at zero bias is given by

$$g_{mo} = -2\, I_{DSS}/V_P \tag{4.2}$$

The mutual conductance g_m at any other value of I_D is given by

$$g_m = g_{mo}\,(1 - V_{GS}/V_P)$$

or $$g_m = g_{mo}\,\sqrt{(I_D/I_{DSS})}. \tag{4.3}$$

4.1.5 *Automatic bias: provision of gate-source bias by means of a source resistor.* The gate may be maintained at a negative potential with respect to the source by the use of the resistor R_S (Fig. 4.4) so dispensing with the need for a second voltage supply.

Set the drain-source voltage V_{DS} at 15 V, vary the resistance R_S in the lead between the source and the gate to a number of values between 200 Ω and 20 kΩ: at each of these values record the drain current I_D. The product $I_D R_S$ gives the bias voltage V_{GS} applied to the gate. Plot a graph of V_{GS} (determined in this manner) against corresponding values of I_D.

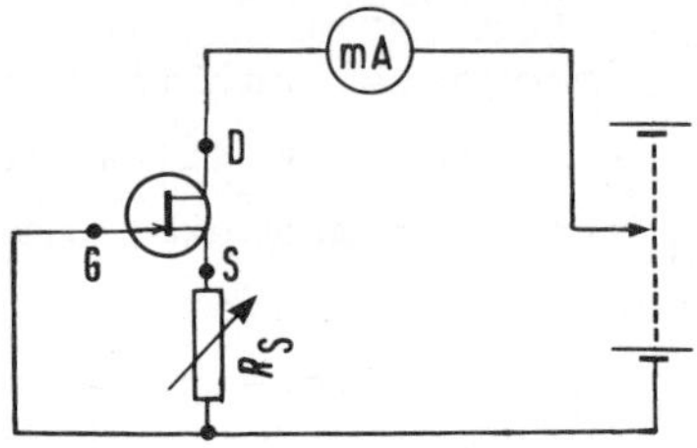

Fig. 4.4 A source resistor R_S to provide gate-source bias V_{GS}; mA is a milliammeter with a fsd of 10 mA and measures I_D.

4.1.6 *Demonstration of the high input resistance of an fet.* With the circuit of Fig. 4.5 set V_{DS} = 15 V, R_s = 1 kΩ and charge the 1 μF capacitor to 2 V by momentarily closing the switch. Record the voltage V across the 1 kΩ resistor at one minute intervals over a period of ten minutes. This voltage will be found to remain constant showing that the drain current I_D is constant, which means that V_{GS} is constant and so negligible charge has left the capacitor via the gate G, which can only be the case if the input resistance of the fet is very large.

It is also of interest here to make some comparisons

(a) connect a reverse-biased silicon diode between the capacitor and the positive line (as shown by dashed lines in Fig. 4.5.)

(b) replace the silicon diode by a 100 MΩ resistor.

In each case (a) and (b) plot the variation of the voltage V with time.

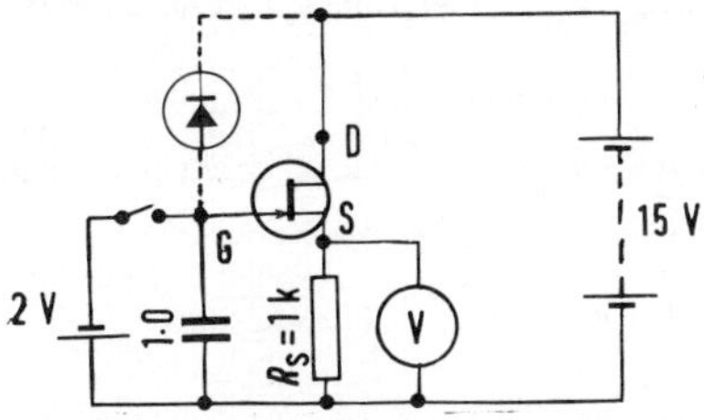

Fig. 4.5 Demonstration of the high input resistance of an fet.

4.1.7 *Further investigation.* Note that the fet and the resistor, R_S in the circuit of Fig. 4.4 form a useful constant current source. Provided that the voltage V_{DS} is greater than that corresponding to the 'knee' of the characteristic (Fig. 4.2b), the drain current I_D is almost independent of the applied voltage.

Connect a load resistance of, say, 500 Ω in series with the milliammeter of Fig. 4.4 set the applied voltage to 10 V and use the variable resistor R_S to set the drain current I_D to 1 mA. Increase the applied voltage in steps between 10 - 20 V, measuring I_D at each setting. Repeat with a drain current I_D of 50 μA.

4.2 *A simple common-source fet amplifier.*
The circuit of Fig. 4.6 is considered first from a numerical point of view. Suppose that the fet provides an I_{DSS} of 8 mA, that V_P is 3 V and that it is planned to operate it with a direct drain current of 3 mA. The

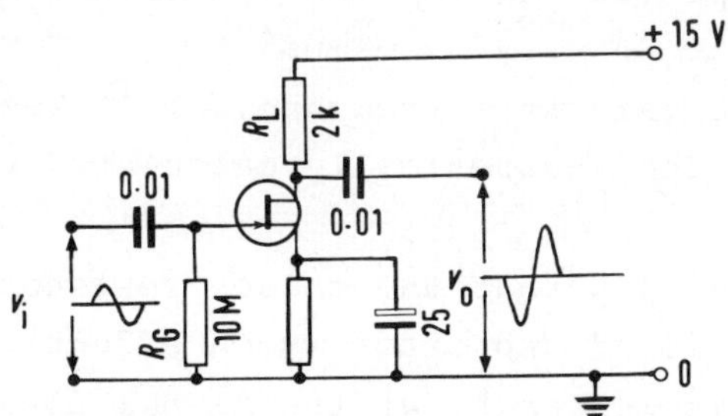

Fig. 4.6 A common-source fet amplifier

gate-source voltage V_{GS} required is given by

$$V_{GS} = 3\left[1 - \surd(3/8)\right] \simeq 1.2\text{V}.$$

To provide this value of V_{GS}, a source resistor R_S is required of magnitude given by

$$R_S = 1.2/\ (3 \times 10^{-3}).$$

The nearest preferred value to this is 390 Ω, so R_S is chosen to be 390 Ω. The mutual conductance at zero bias is g_{mo} given by

$$g_{mo} = \frac{2 \times 8}{3} = 5.3 \text{ mA V}^{-1}$$

At a drain current I_D = 3 mA

$$g_m = 5.3 \surd\ (3/8) \simeq 3.2 \text{ mA V}^{-1}$$

With a load resistance R_L of 2 kΩ, the voltage gain of this amplifier (Fig. 4.6) is

$$A_V = V_o/V_i = R_L\ i_D/V_{GS} = R_L\ g_m = 2 \times 3.2$$
$$= 6.4$$

4.2.1 *The voltage gain of a fet amplifier.* V_{DS} is set at 15 V, and an input signal of, say, 100 mV peak-to-peak from a known variable frequency (say 10 Hz to 20 Hz) signal generator is used. Observation by a calibrated double-beam oscilloscope enables determinations to be made of (a) the phase change introduced by this amplifier and (b) the voltage gain v_{out}/v_{in} at various frequencies over the range 10 Hz to 20 kHz. From the results of (b) a graph can be plotted of the voltage gain against the frequency.

The voltage gain at a frequency of 10 kHz should be compared with the theoretically calculated value (see above).

Note in Fig. 4.6 that the source resistor R_S is by passed with a 25 μF capacitor which effectively provides a short circuit across R_S at the signal frequency.

4.3 *Sinusoidal waveform generators based on field-effect transistors*

In Sections 3.9.1, 3.9.2 and 3.9.3 are descriptions of sinusoidal oscillators based on bipolar transistors which make use respectively of phase-shift, quartz crystal control and the Wien bridge principle. In each of these cases the active device could be a field-effect transistor instead of a bipolar but the supply voltage and passive component values would both need to be altered. Moreover the gain obtainable is usually insufficient to enable a phase-shift oscillator to be based easily on a field-effect transistor.

4.4 *Multivibrators utilizing fets*

The three basic classes of multivibrator were introduced in Section 3.10.

4.4.1 *A Free-running multivibrator based on the use of n-channel fets.* The circuit of Fig. 4.7 offers the advantage that, compared with the bipolar transistor, the high input resistance of the fets enables relatively long time constants to be obtained without the need for large capacitor values.

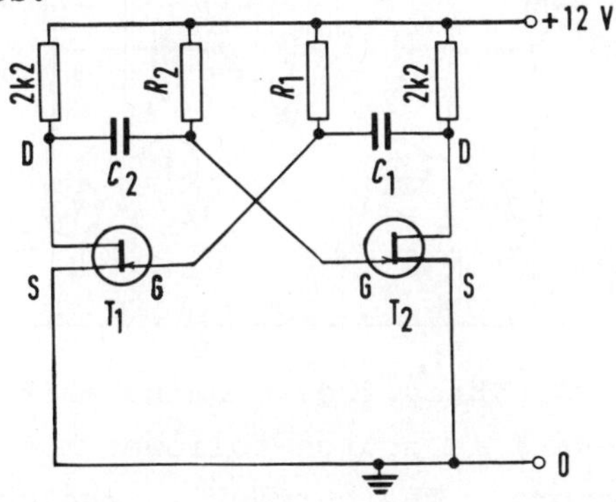

Fig. 4.7 A free-running multivibrator based on the use of n-channel fets.

With the circuit of Fig. 4.7 in which the n-channel fets are both of the BFW 61 type in a symmetrical configuration, the relationship

$$T = 1.38\ RC$$

can be verified experimentally, where T is the time period of the oscillation, $R = R_1 = R_2$ and $C = C_1 = C_2$.

Another experiment which can be undertaken with this circuit is to examine the very low frequency behaviour with capacitance values of 1 μF. Light-emitting diodes (LEDs) are used in place of the 2.2 kΩ resistors to detect the conducting state of the fet (visible light is emitted when the LED is forward-biased and carrying a current between 5 and 10 mA). A 1 kΩ resistance should be used in series with each LED.

4.4.2 *A hybrid free-running multivibrator.* In the circuit of Fig. 4.8 the bipolar transistors provide good switching characteristics, giving an output voltage with a peak-to-peak value nearly equal to the supply voltage, whereas the fets provide the high input impedance necessary for reliable low frequency operation. Hence the advantage of each type of transistor is utilized.

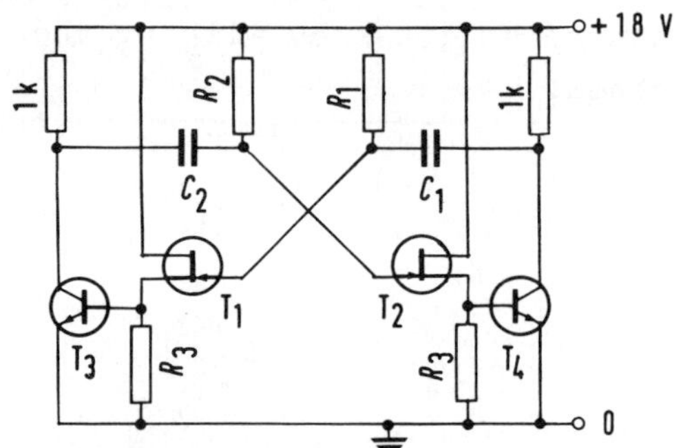

Fig. 4.8 A hybrid free-running multivibrator.

The fets operate in source-follower mode to drive the bipolar transistors. The current in the collector leads of the bipolar transistors is much greater than could be available from the fet version of the multivibrator, and could be used to operate small electromagnetic devices.

4.4.3 *A voltage-to-frequency converter based on a free-running multivibrator utilizing field-effect transistors.* The time period (T) of a multivibrator can be changed without altering the actual component values, but by making use of a variable voltage supply, as shown for a bipolar transistor-based multivibrator in Fig. 3.21.

A voltage-to-frequency converter of superior performance to that of Fig. 3.21 is based on the use of n-channel fets (Fig. 4.9) in which each of the resistors R_1 and R_2 of Fig. 4.8 is replaced with a bipolar transistor acting as a constant current device. The magnitudes of the currents which charge the two capacitors C (both equal to 1 nF in Fig. 4.9) are constant and controlled by the emitter-base voltage V_{eb} (adjusted to between 0.2 and 0.8 V for the bipolar transistors in Fig. 4.9). This provides a circuit with very good linearity between the time period (T) and voltage V_{eb} (adjusted to experiment on such a circuit). The frequency of the multivibrator (actually the reciprocal of the time of one cycle of the output waveform) is measured by a cro having a calibrated time-base; this frequency should extend over at least five decades.

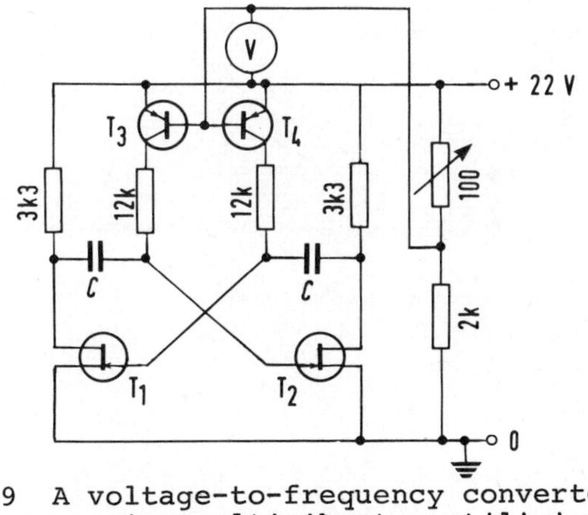

Fig. 4.9 A voltage-to-frequency converter based on a free-running multivibrator utilizing field-effect transistors.

It is interesting to note how very temperature sensitive this circuit is. The BCY70 bipolar transistors of Fig. 4.9 are mounted on heat sinks but even so the frequency is affected by touching the heat sink with a finger. The heat sensitivity of this circuit means that it can be used to measure temperature in terms of a change in frequency.

4.4.4 *A monostable multivibrator with a pulse width determined by a fet-based constant current source.* The circuit of Fig. 3.23a for a monostable multivibrator utilizing bipolar transistors is set up with the discharge resistor *R* replaced by a fet constant current arrangement as in Fig. 4.10. The waveform of the output voltage at the base of the bipolar transistor T_2 is displayed on a cro. To change the pulse width, the resistance (10 kΩ max) in the source lead of the fet is altered.

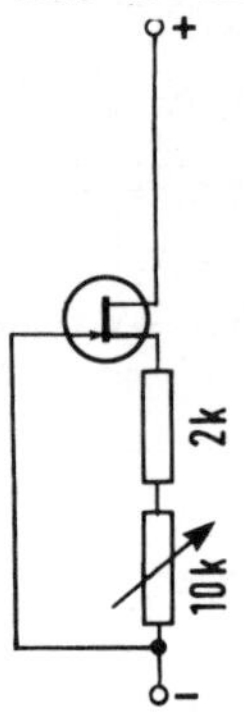

Fig. 4.10 A fet constant current source to be used in place of the resistor *R* in Fig. 3.23(a) to provide a monostable multivibrator with a determinable pulse width.

CHAPTER FIVE

Unijunction transistors; silicon controlled rectifiers: characteristics and applications

5.1 *Unijunction transistors (UJTs)*

The unijunction transistor (Fig. 5.1) consists of a rod of n-type silicon to the ends of which ohmic (non-rectifying) contacts B_1 and B_2 are made. The resistance of the silicon rod, called the interbase resistance, is usually between 5 kΩ and 10 kΩ. On one side of this silicon rod is formed a p-n junction by alloying a wire of aluminium (trivalent so producing p-type material); this junction is usually located near to the base B_2. An emitter electrode E is brought out from the p-type material; this emitter divides the interbase resistance in the ratio R_{B2} to R_{B1}.

The voltage of the emitter relative to base B_1 is a constant fraction of the applied interbase voltage V_{BB}. This constant fraction η - called the intrinsic stand-off ratio - is readily seen to be decided by

$$\eta = R_B / (R_{B1} + R_{B2}) \qquad (5.1)$$

The voltage relative to B_1 required to forward bias the emitter junction is called the emitter peak-point voltage V_P. While the emitter voltage V_E is less than V_P, the emitter junction is reverse-biased and only leakage current of a few microamperes will flow. The peak point voltage V_P is related to the interbase voltage V_{BB} by the equation

$$V_P = \eta V_{BB} + V_D \qquad (5.2)$$

where V_D is the equivalent diode voltage across the emitter junction. When the junction is forward biased, positive holes are injected into the n-type silicon.

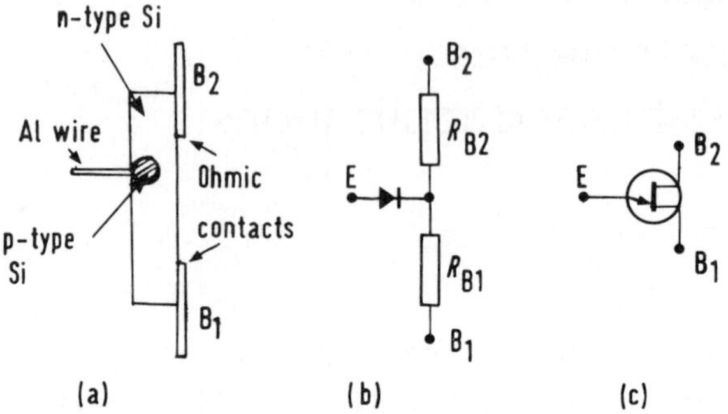

Fig. 5.1 The unijunction transistor: (a) structure; (b) equivalent circuit; (c) circuit symbol.

Because B_2 is usually maintained at a positive potential with respect to B_1, these holes are swept into the base B_1 region and R_{B1} decreases rapidly in magnitude. The pd between the emitter and base B_1 decreases so that the emitter junction is further forward biased and this causes the emitter current to increase. Because the emitter current I_E increases while the emitter voltage V_E decreases, the device exhibits a negative resistance region in the characteristic emitter current against emitter voltage. This leads to the use of the UJT in a simple relaxation oscillator circuit (Section 5.2).

5.1.1 *The intrinsic stand-off ratio.* Using the circuit of Fig. 5.2 the emitter current has to be limited with a series 2 kΩ resistor, otherwise the junction will be damaged. For various values of V_{BB}, the measured voltage V_E is increased slowly. When V_E equals V_P, the voltage on the emitter drops suddenly as current flows into the silicon rod via the forward biased emitter junction: the values of V_P corresponding to various measured values of V_{BB} are recorded, where V_P should be measured a number of times for each selected value of V_{BB} up to 25 V. From Equation (5.2)

$$V_P + \eta V_{BB} = V_D$$

so that a plot of V_P against V_{BB} enables values of both η and V_D to be determined

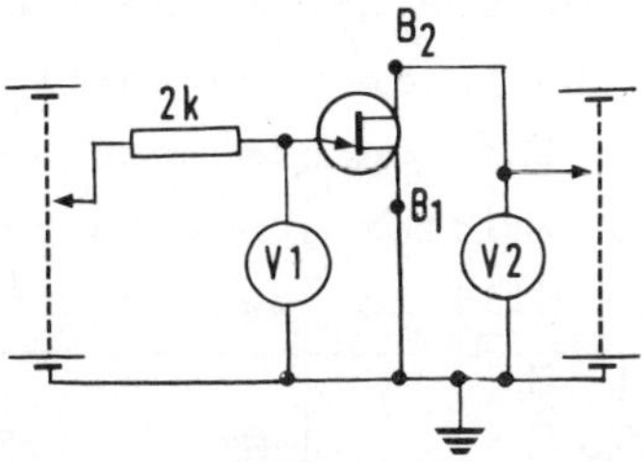

Fig. 5.2 Determination of the intrinsic stand-off ratio of a ujt. V1 and V2 are high resistance voltmeters: VI measures V_E; V2 measures V_{BB}.

5.2 *Relaxation oscillators*

In a relaxation oscillator a capacitor C is charged through a resistor R and is then rapidly discharged so as to return to its initial state when repetition of this process takes place.

5.2.1 *A relaxation oscillator based on a unijunction transistor.*

A typical circuit is given in Fig. 5.3. This very simple oscillator circuit provides an output of non-sinusoidal waveform. The voltage waveform (a sawtooth) across the 0.1 μF capacitor C_1 and the voltage waveform (a series of positive-going pulses) across the 56 Ω resistor are observed and recorded simultaneously by the use of a double-beam cro. It is a simple matter to establish that the oscillator frequency is independent of the voltage applied across the oscillator for given values of R and C_1.

The resistance R (conveniently in the form of a resistance box of maximum value 10^5 Ω approximately) is varied between 5 kΩ and 8 kΩ; at each setting of R the frequency of the output pulses across the 56 Ω resistor is measured in terms of the calibrated time-base of the cro. The frequency f of this oscillator can be shown to be given approximately by

$$f = 1/|RC \log_e (1/1-\eta)|$$

A plot of f against $1/R$ enables a value for the intrinsic stand-off ratio η for the ujt to be found.

The value of R can be reduced until the oscillator ceases to function; this occurs when too much current is passing through the emitter junction so that the unijunction transistor will not switch off.

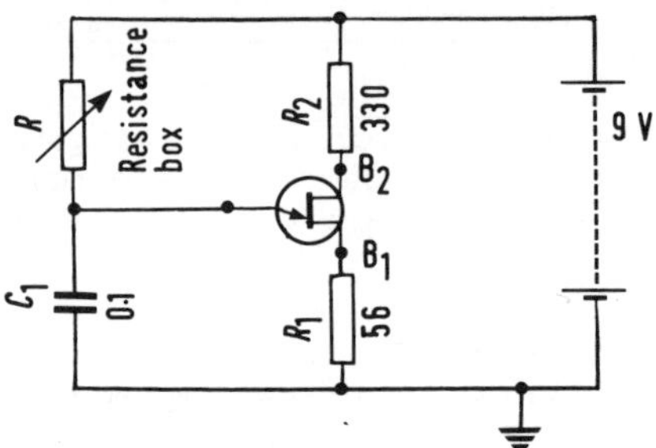

Fig. 5.3 A relaxation oscillator based on a unijunction transistor.

5.3 *A staircase generator or frequency divider based on a unijunction transistor*

In the circuit of Fig. 5.4 the unijunction transistor UJT_1 forms part of a relaxation oscillator of which the frequency is governed by the variable resistor R_1. Each time the 0.1 μF capacitor is discharged, a negative pulse appears across R_2 (a variable resistor of maximum value 50 Ω). The bipolar transistor T_1 is normally non-conducting because its emitter and base are at the same potential. When the negative pulse arrives, the emitter base junction of T_1 is forward-biased so T_1 conducts for the duration of the pulse. The tapping point on the potentiometer R_2 will determine the magnitude of the forward bias on T_1 and hence the magnitude of the current pulse which charges the capacitor C_1. Each pulse increases the charge in C_1 in well-defined steps until

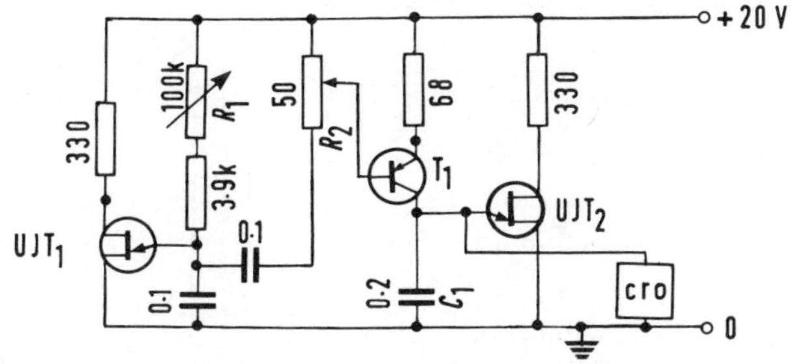

Fig. 5.4 A staircase generator based on a unijunction transistor.

the voltage across C_1 reaches the peak point value of UJT_2 which will then discharge the capacitor C_1.

The output waveform is recorded by a cro connected as shown in Fig. 5.4. The manner in which the resistance controls the number of cycles occurring per second should be noted, while the value of R_2 determines the number of steps in each cycle. It is a simple matter to arrange for say, one pulse at the output for every 20 pulses at the input.

The number of such divider stages which can be connected in cascade will be determined by the leakage of the capacitors and the UJTs. Note that the negative-going pulse serves to open the gate T_1 for a finite time.

5.4 *Programmable unijunction transistors (puts)*

These are planar, passivated, four-layer (p-n-p-n) devices with three terminals: the anode; the anode-gate; and the cathode. Compared with the ujt, the put offers the chief advantages of (i) low leakage current, (ii) low peak point current I_P and (iii) the facility of being able to arrange the values of the intrinsic stand-off ratio η and the interbase resistance R_{BB} by appropriate choice of resistors external to the put. The circuit symbol and structure of the put are shown in Fig. 5.5

The put is used in trigger circuits for silicon controlled rectifiers, in pulse circuits, timing circuits, sensing circuits and oscillators.

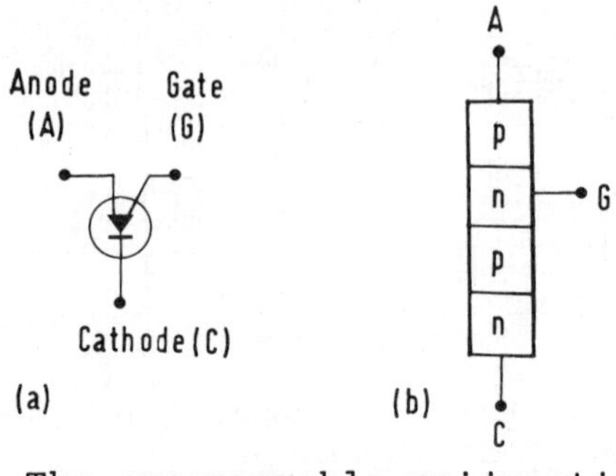

Fig. 5.5 The programmable unijunction transistor (put): (a) the circuit symbol; (b) the structure.

5.4.1 *Selection of the value of the intrinsic stand-off ratio with a put.* Making use of the circuit of Fig. 5.6, with the supply V_{CC} at 20 V and $R_1 = R_2 = 5$ kΩ, increase the voltage on the anode until it reaches the peak point value V_P; note that current does not flow through the put until the anode voltage reaches this critical value. Repeat for other values of R_1 and R_2 with R_1 not necessarily equal to R_2 and where $(R_1 + R_2)$ always exceeds 10 kΩ. The results are conveniently entered into a table with the following column headings V_{CC}, R_1, R_2, V_P, $\eta = V_P/V_{CC}$, $\eta = R_2/(R_1 + R_2)$.

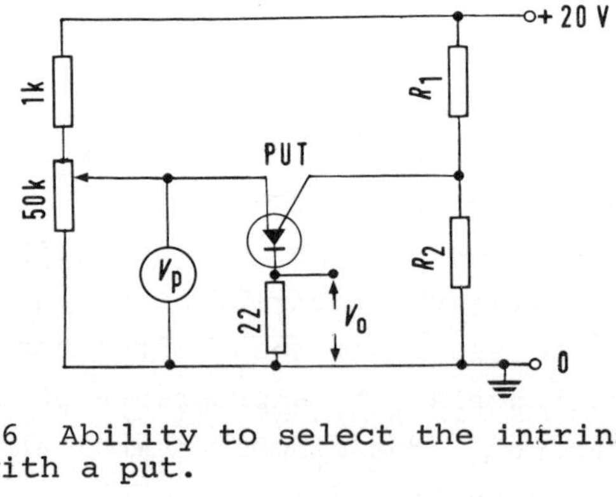

Fig. 5.6 Ability to select the intrinsic stand-off ratio with a put.

5.5 *A relaxation oscillator based on a put.*

Set up the circuit shown in Fig. 5.7. By means of a cro

observe and record the output pulses V_0 across the 22 Ω resistor. Measure the peak height of these pulses and estimate their rise time. Alter the value of the variable resistor R and measure the frequency of oscillation (in terms of the calibrated time-base of the cro) given by

$$f = 1/(RC \log_e (1/1-\eta))$$

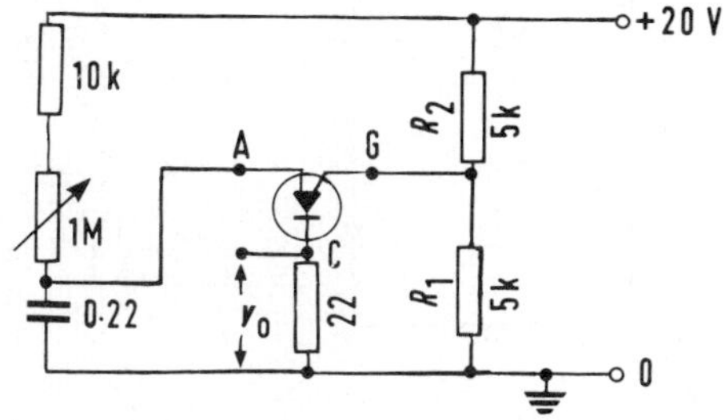

Fig. 5.7 A relaxation oscillator based on a put.

By drawing an appropriate graph (based on the equation) and from the results obtained, deduce values of η and compare them with those calculated from the values of R_1 and R_2 used.

5.6 *Silicon controlled rectifiers (scrs)*

The scr is similar to a conventional silicon rectifier but has an additional third electrode called the gate (Fig. 5.8a).

The flow of current through an scr is zero when it is reverse-biased, as in the case of a conventional diode. However, even when forward-biased it will not conduct, unless it is triggered into conduction by means of a signal applied to the gate electrode. The firing signal has to involve power so that, although it must make the gate positive in potential with respect to the cathode, essentially it needs to be a pulse of current. Once this pulse has triggered the scr into a conducting state, the gate no longer exerts any control. It follows that the power involved in the triggering pulse is wasted - so it is kept as small as possible. To turn off the scr, either

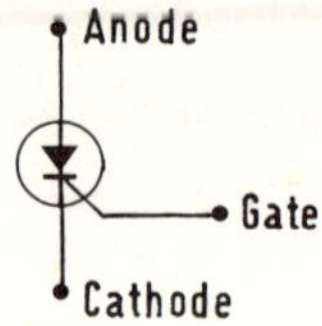

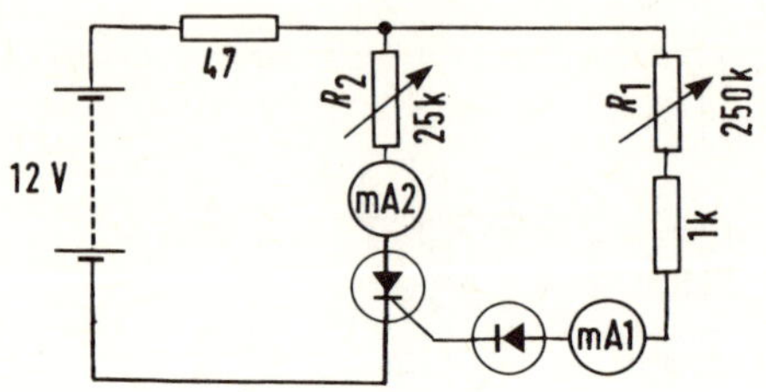

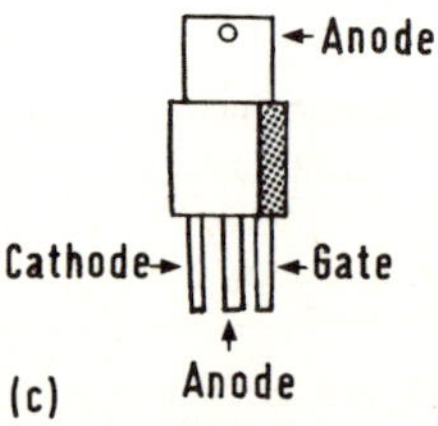

Fig. 5.8 The silicon-controlled rectifier (scr): (a) circuit symbol; (b) determination by experiment of the gate current required to turn on an scr, and of the holding current; (c) schematic diagram of the ICR10 silicon-controlled rectifier.

a reverse bias is applied for an instant across the anode-cathode or the current through the scr is reduced to below a critical value known as the 'holding current'.

The scr is one of the most sensitive control devices available because the consumption of power in the firing pulse applied to the gate circuit usually requires a power dissipation of only a few tens of milliwatts, whereas the power switched in the anode circuit may well be several kilowatts: a power gain of 10^5 or more can thus be provided.

Note. Although not relevant to the experiments described,

an scr can also be switched into a conducting state (turned on) by a very large rate of rise of the forward applied voltage across anode and cathode, ie by a very large *dV/dt*.

5.6.1 *To determine the gate current required to turn on an scr.* The scr (an IRC10 is a convenient model) is plastic encapsulated and mounted on a small aluminium heat sink. This scr is designed to carry a maximum forward current of 1 A and to withstand a maximum reverse pd of 100 V.

The circuit of Fig. 5.8b is used to find experimentally the gate current required to turn on the ICR10: in this circuit, the meter mA2 in the anode lead has a fsd of 25 mA whereas the meter mA1 in the gate lead has a fsd of 1 mA.

The resistance of the variable resistor R_1 is reduced progressively from its maximum value of 250 kΩ so as to increase progressively the gate current (as recorded by mA1) to the value at which the scr turns on, as is indicated by the current shown by mA2. On disconnecting the firing circuit it is noted that the scr is still conducting provided that the current through it is sufficiently large. The resistance R_2 in the anode circuit is increased until the scr switches off (the reading of mA2 $\rightarrow$ 0): the current at which this occurs is the holding current. The experiment is repeated a few times so as to find a mean value of the gate current required for switching, and the holding current.

5.7 *Phase control by means of silicon controlled rectifiers*

Phase control is the process of rapid on-off switching which connects an ac supply to a load for a controlled fraction of each cycle, where this fraction is determined by phase. This forms a very efficient way of controlling the supply of ac power to lamps, heaters, motors etc., and the silicon-controlled rectifier (scr) is an ideal device to use for such phase control. The control is achieved by governing the phase angle of the ac waveform at which the scr is triggered. The scr will switch from a non-conducting to a conducting state in 5 μs approximately and conduct for the remainder of the half-cycle.

5.7.1 *Half-wave phase control with a phase angle between 0 and π/2.* The circuit of Fig. 5.9a is set up. In this circuit a 12 Vrms ac supply is used at 50 Hz from the secondary of a mains step-down transformer. The magnitude of the variable resistor R_1 (maximum: 250 kΩ) in the gate circuit is reduced whilst the voltage waveform across the 12 V, 5W electric lamp bulbs is recorded by means of a cro. The form of this voltage waveform is shown in Fig. 5.9b.

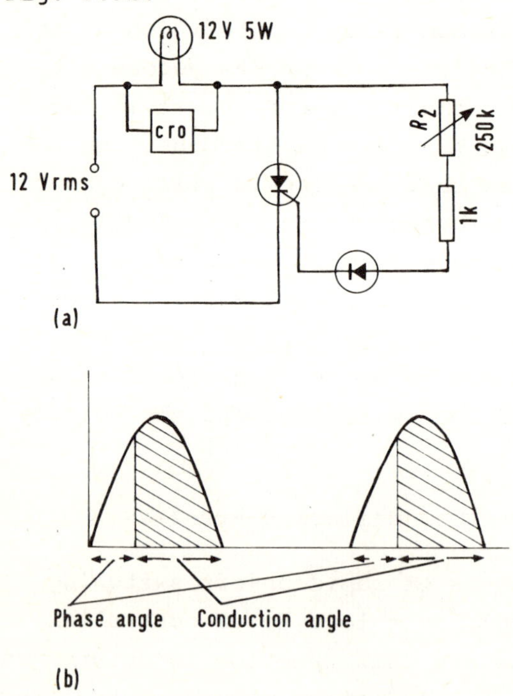

Fig. 5.9 (a) Circuit diagram for half-wave phase control with phase angle between 0 and π/2. (b) The voltage waveform pertinent to the operation of this circuit.

5.7.2 *Half-wave phase control with a phase angle between 0 and π.* The circuit of Fig. 5.10 is used. The value of the capacitor *C* can be increased by connecting further capacitors in parallel with it. As in Section 5.7.1, the voltage waveform across the 12 V, 5 W electric lamp bulb is recorded.

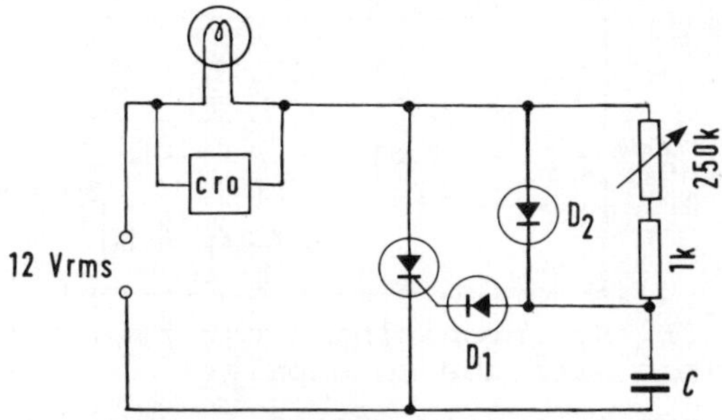

Fig. 5.10 Half-wave phase control with phase angle between 0 and π .

5.7.3 *Questions.*

(a) In the experiment described in Section 5.7.1 why can one only vary the phase angle at which the scr is switched on between 0 and π/2?

(b) What is the function of the diode D_2 in the circuit diagram of Fig. 5.10? Why does the introduction of the capacitor *C* enable the phase angle to exceed π/2?

5.8 *Phase control by means of an scr fired by pulses from a ujt circuit*

The sharp positive-going pulses available from a unijunction transistor relaxation oscillator are convenient for triggering an scr. The pulse from such an oscillator must be synchronized with the ac supply. To achieve this a number of methods are possible. In Fig. 5.11 a unijunction transistor (ujt) oscillator is shown which is supplied with a full-wave rectified signal of which the peaks of the voltage waveform are removed by means of a Zener diode. At the end of each half-cycle, the voltage across the ujt will fall to zero so the capacitor C_1 will be discharged through the emitter of the ujt. The time constant of the network R_1C_1 will determine the time taken for this emitter to come up to the peak-point voltage and so produce the first pulse which will trigger the scr.

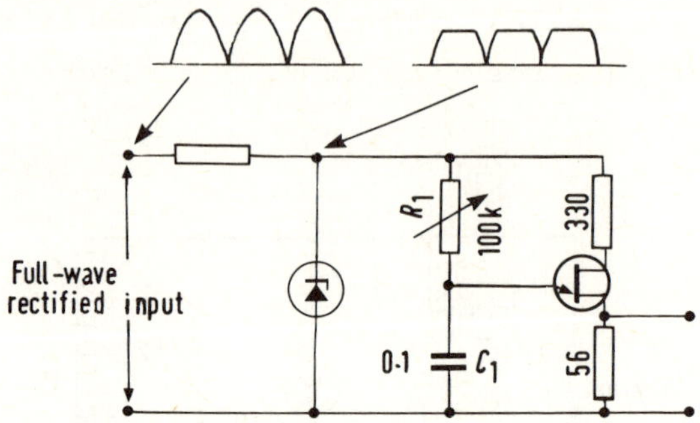

Fig. 5.11 Synchronizing a ujt relaxation oscillator with the ac supply.

A similar synchronizing method is used in the circuit of Fig. 5.12 except that, while a second ujt resets the voltage across the capacitor C_1 at the end of each half-cycle, the timing circuit is operated from a stable dc supply.

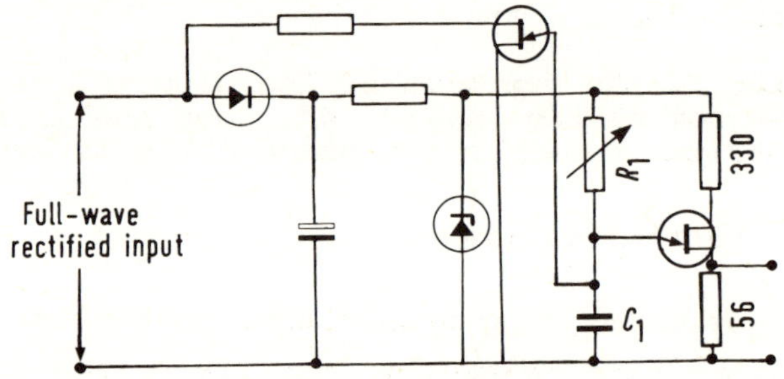

Fig. 5.12 An alternative method of synchronization to that of the circuit in Fig. 5.11.

5.8.1 *Phase control utilizing two scrs.* Set up the ujt oscillator part (left-hand side) of the circuit shown in Fig. 5.13. Observe with a cro the voltage waveform across the 56 Ω resistor as the value is varied of the resistor of 50 kΩ (max). Notice the fast leading edges of these pulses and record their height. Link the two halves of the circuit so that the positive-going pulses produced

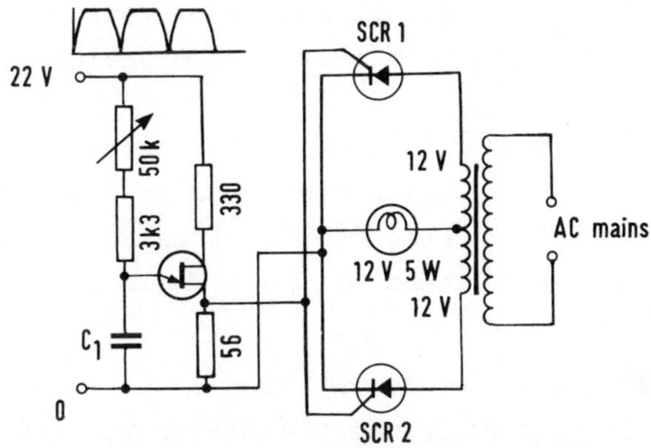

Fig. 5.13 Phase control utilizing two scrs.

across the 56 Ω resistor are applied between the gates and cathodes of the two scrs 1 and 2. The heights of these pulses will be observed to decrease when these connections are made. Why? Whichever scr is forward biased when the gate pulse arrives will be triggered into conduction and will conduct for the remainder of that half cycle. Observe the voltage waveform across the load (a 12 V, 5 W filament lamp) as the value of the resistor of maximum value 50 kΩ is varied. Record the waveform with (a) the filament bright and (b) a restricted current through the filament so that it is just visible.

5.8.2 *Alternative control of the phase angle.* In the circuit of Fig. 5.13 the value of the variable resistor of 50 kΩ max is set so that triggering occurs very early in the cycle, and a decade resistance box is connected across the capacitor C_1 which will serve to shunt some of the charging current to C_1 in order to increase the time taken for the emitter of the ujt to reach its peak point value V_P. If the value of this shunting resistance is very large, the current drain from the capacitor C_1 is negligible. As this resistance is decreased the phase angle is increased until, at some low value of this resistance the emitter of the ujt does not reach peak point voltage during the cycle so that no firing pulse occurs. Measure the phase angle for various values of this shunt resistance and plot a graph of phase angle against shunt resistance.

Replace the decade resistance box across C_1 by an n-channel field effect transistor used as a voltage controlled resistor. The firing unit is then voltage controlled so a graph can be plotted of phase angle against the fet gate-source voltage.

5.8.3 *Utilizing a ujt firing circuit with simple negative feedback.* In Fig. 5.14 the thermistor Th in series with the variable resistor R_2 across the capacitor C serves to control the firing angle of the scrs 1 and 2 in Fig. 5.13, if the left-hand part of the circuit of Fig. 5.13 is replaced by the circuit of Fig. 5.14. The correct choice of this thermistor will ensure that when cold its resistance will provide negligible shunting so that the scr will fire early in the cycle and the maximum power is delivered to the load. If the thermistor bead (having a negative temperature coefficient) is placed in contact with the glass bulb of the filament lamp (12 V, 5 W of Fig 5.13) its temperature will rise so that its resistance will fall and the lamp will dim.

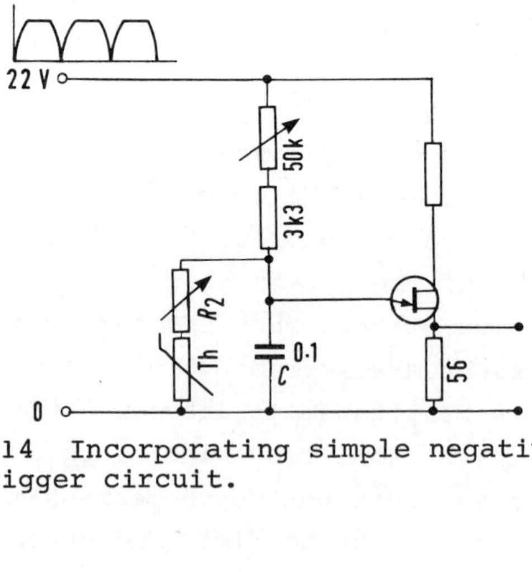

Fig. 5.14 Incorporating simple negative feedback in a trigger circuit.

5.8.4 *Further investigations.*

(a) Modify the circuit of the feedback system of Fig. 5.14 so that a cadmium sulphide photoconductive cell is used to maintain the illumination (at a point close to the filament lamp) at some predetermined level.

(b) A number of phase control modules are available commercially. An example is the MR16 unit (Reyrolle Parsons) in which the phase angle is controlled by a steady voltage of between 0 and 3 V. The experimental study of phase-control can be readily extended with this unit.

(c) Domestic light dimming circuits based on the use of a TRIAC (equivalent to two SCRs connected in inverse parallel) merit experimental study. It is convenient to operate them from the secondary of a transformer so as to obtain electrical isolation from the ac mains supply.

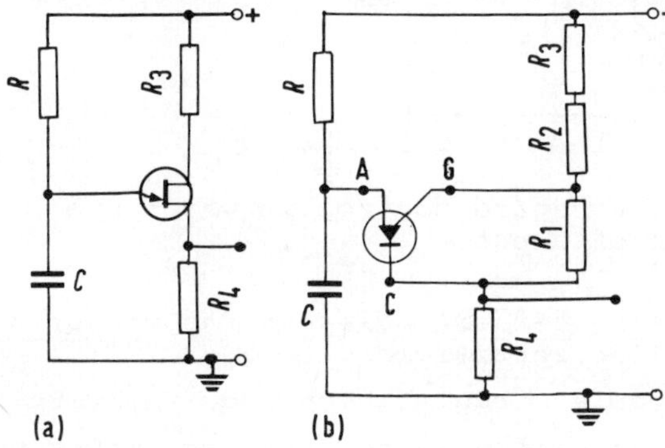

Fig. 5.15 Comparison between basic relaxation oscillators using (a) a ujt and (b) a put.

5.9 *Phase control by means of a put*

When a conventional ujt relaxation oscillator is used to trigger an scr the resistor R_4 (Fig. 5.15a) serves to by-pass the standing current which flows through R_{BB} which would otherwise trigger the scr. When a put is used instead, there is no standing current, so the cathode can be connected directly to the gate of the scr (Fig. 5.16). Whereas in the ujt circuit (Fig. 5.15a) the resistor R_3 is normally used to stabilize the ujt working point, this is not necessary in the put circuit because R_{BB} consists of fixed linear resistors.

With the circuit arrangement of Fig. 5.16 use the output pulses from the put to trigger the scr, A full or half-wave rectified supply must be used so that the voltage across the scr drops to zero at the end of each half-cycle and therefore the scr will switch off. Record by a cro the voltage waveform across the filament lamp when it is glowing brightly and also when it is just visible.

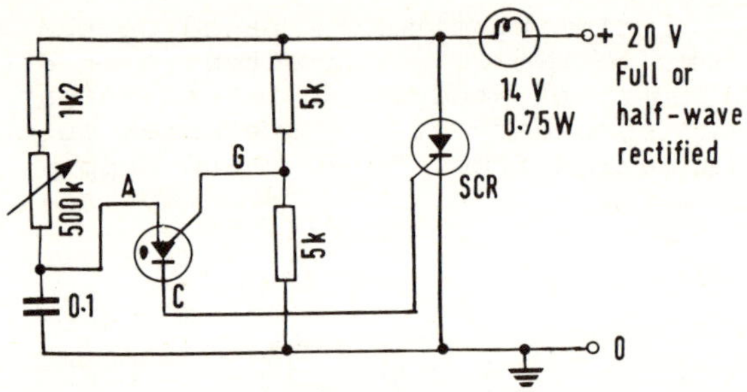

Fig. 5.16 Phase control in which use is made of a put.

5.10 *A bistable circuit based on the use of silicon controlled rectifiers*

A simple bistable circuit which requires very few components and can switch large currents, makes use of two silicon controlled rectifiers (scrs). See Fig. 5.17. When the steady voltage (12 V supply) is on, neither scr will conduct although each of them is forward-biased. A current pulse to the gate electrode of scr 2 (from the control 'box') which makes this gate positive with respect to the cathode will switch this scr into a conducting state, so that current will flow through the load resistance R_L. The commutating capacitor C is thereby charged with the polarity shown in Fig. 5.17. Switching in scr 1 by applying a current pulse to its gate will connect the charged capacitor C directly across scr 2 which, for an instant, is thereby reverse-biased and so switches off.

With scr 1 conducting and scr 2 not conducting, the capacitor *C* is charged with the opposite polarity from that shown in Fig. 5.17 and so scr 1 is switched off when the next firing pulse is applied to scr 2. For effective switching a minimum value of the capacitance *C* is required; the greater the load current, the larger the capacitance needed for reliable switching.

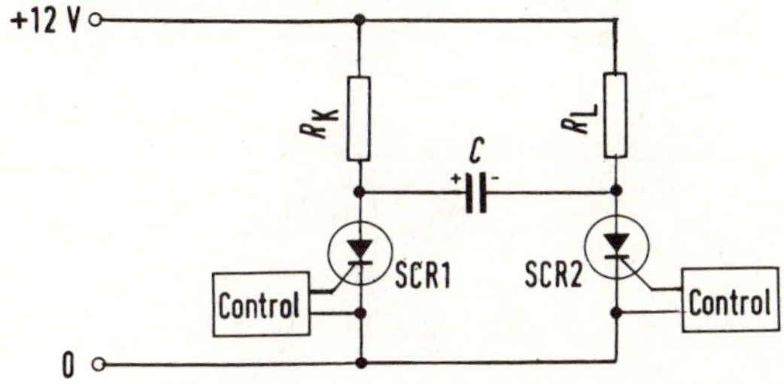

Fig. 5.17 A simple bistable circuit based on the use of silicon-controlled rectifiers.

A first experiment is based on the circuit of Fig. 5.18a. When the switch S_2 is closed a current pulse passes through the 680 Ω resistor and the gate-cathode circuit of scr 2 so that the lamp (forming the load resistance R_L, cf Fig. 5.17) is lit and the commutating capacitor (1 μF) charged. Closing the switch S_1 will provide a current pulse to the gate of scr 1 which is thereby triggered into the conducting state and simultaneously scr 2 is switched off. The flow of a current pulse through one or other of the 680 Ω resistors when the appropriate switch (S_1 or S_2) is closed can be observed with a cro across the resistor to show the voltage waveform.

The 1 MΩ resistor across the 0.47 μF capacitor (in both cases) serves to discharge the capacitor after the switch is opened.

In a second experiment the bistable circuit based on two scrs is switched by means of pulses from an astable multivibrator which is designed, by the use of large

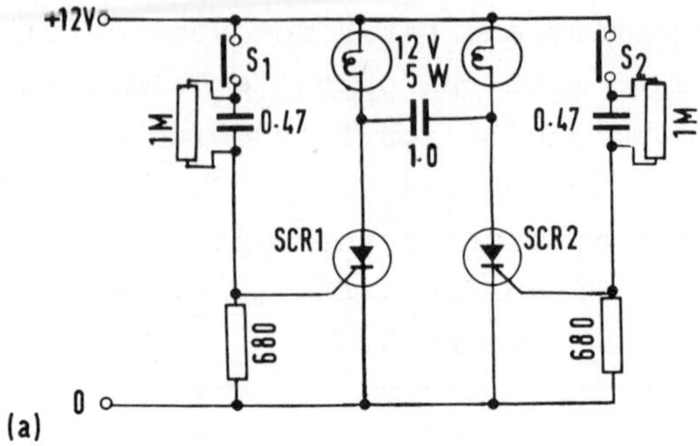

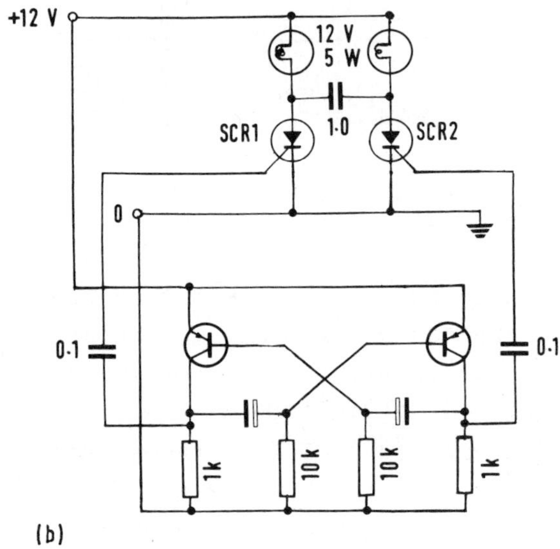

Fig. 5.18 Experiments with a bistable circuit based on scrs: (a) with hand-operated switches; (b) with switching by means of a low-frequency free-running multivibrator.

electrolytic capacitors (each of 100 μF) to oscillate with a period of approximately 2 s (Fig. 5.18b). When either bipolar transistor conducts, a positive-going pulse will appear across the 1 kΩ resistor in its collector lead. These pulses are communicated via the 0.1 μF capacitors to the respective gates of the scrs.

The switching action originating from the astable multivibrator has been made so slow that the behaviour of the bistable circuit can be observed readily with filament lamps as loads. Of course, both the astable multivibrators and the bistable scr circuits could be designed to operate at much high frequencies.

CHAPTER SIX

More complex amplifiers and some applications

Two classes of more complex amplifier are considered, where 'more complex' as opposed to 'simple' implies amplifiers comprising more than a single active device such as a bipolar transistor or a field-effect transistor. The first class of more complex amplifier is the differential or difference amplifier. The second class is that of the multi-stage amplifiers which comprise a set of stages 1, 2,3 and 4 (more than 4 is very unusual), where the output from stage 1 is coupled to the input to stage 2 and so on. Each stage is a simple single-stage amplifier based on a single active device or may be a differential amplifier which is based on two active devices which, however, are not in cascade.

6.1 *Differential or difference amplifiers*

These are particularly useful. A differential amplifier is frequently based on two jgfets T_1 and T_2 which have a common supply voltage + V_{DD} and a common source resistor R_S. In the much-used case of a symmetrical differential amplifier (Fig. 6.1) the drain-load resistances are also equal, both being R_D, the input circuits are identical and the two jgfets are identical: to ensure the last of these the jgfets are usually fabricated on the same n-type silicon slice and they are diffusion isolated giving the so-called monolithic dual fet construction (eg type

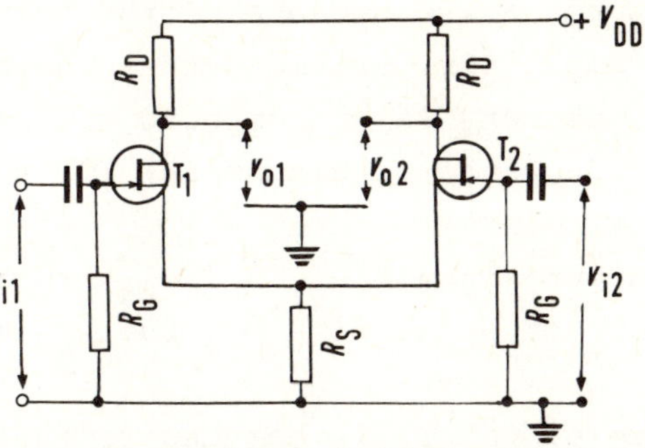

Fig. 6.1 A symmetrical differential amplifier based on n-channel jgfets (symmetry implies that jgfet T_1 is identical with jgfet T_2)

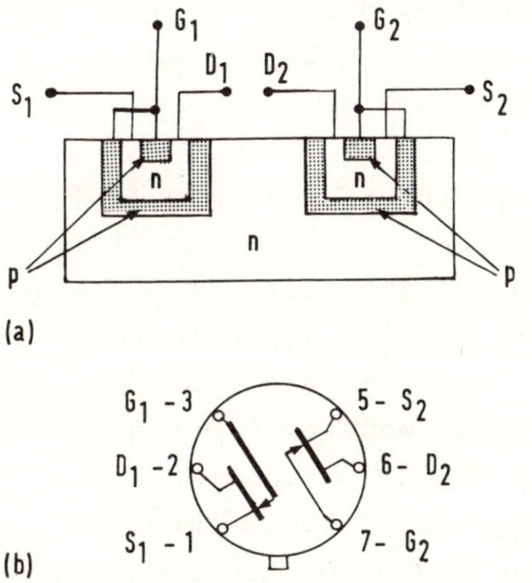

Fig. 6.2 (a) Structure of monolithic dual fets (n-channel jgfets) (type 2N 3955). (b) pin configuration of 2N 3955, viewed from underside; pins 4 and 8 missing, substrate connected to case.

2N 3955 of Fig. 6.2).

The noteworthy feature of this kind of amplifier is that two input signals v_{i1} and v_{i2} can be applied, two corresponding output signals v_{01} and v_{02} are obtained and the magnitude of each of these output voltages depends on the difference between the two input signals so that both v_{01} and v_{02} depend upon $(v_{i1} \sim v_{i2})$.

These features arise because of the symmetry of the differential amplifier, provided that the common source resistance R_S is very large compared with $1/g_m$, g_m being the mutual conductance of either fet (T_1 or T_2, which are identical).

The two output signals v_{01} and v_{02} are in antiphase; v_{01} is 180° out of phase with v_{i1} but v_{02} is in phase with v_{i1}.

If the gate of fet T_2 is earthed so that $v_{i2} = 0$, ie only one input v_{i1} is present, the two outputs are of equal amplitude and in antiphase: this gives the phase-splitter or so-called *paraphase amplifier*.

If the two input signals V_{i1} and V_{i2} are alternating, equal in amplitude and in phase, the alternating output voltage at each output is zero.

When the two input signals are of different amplitudes the output is proportional to the difference between them.

The differential amplifier is sometimes used in such measuring instruments as ac voltmeters and cros where a response extending from zero frequency (dc) to many megahertz (MHz) is required. Differential amplifiers are often the basic stages of an operational amplifier (opamp) usually in integrated circuit form.

In the linear region of the ideal differential amplifier the single output voltage V_0 is given by

$$V_0 = A_d\,(V_{i1} - V_{i2}) \qquad (6.1)$$

where A_d is the voltage gain of the differential amplifier. Any signal which is common to both inputs will clearly have no effect on the output voltage.

In a practical differential amplifier the output depends not only on the difference signal V_d $(= V_{i1} - V_{i2})$ but also on the common mode signal $V_c\ |= \frac{1}{2}\,(V_{i1} + V_{i2})|$.

Thus, if V_{i1} = 1 mV and V_{i2} = 1 mV, the output will not be exactly the same as if V_{i1} = + 449 mV and V_{i2} = 451 mV, even though v_d is 2mV in both cases. The output from such a differential amplifier as that of Fig. 6.3 can be expressed as a linear combination of the two input voltages, ie

$$v_0 = A_1 v_{i1} + A_2 v_{i2} \tag{6.2}$$

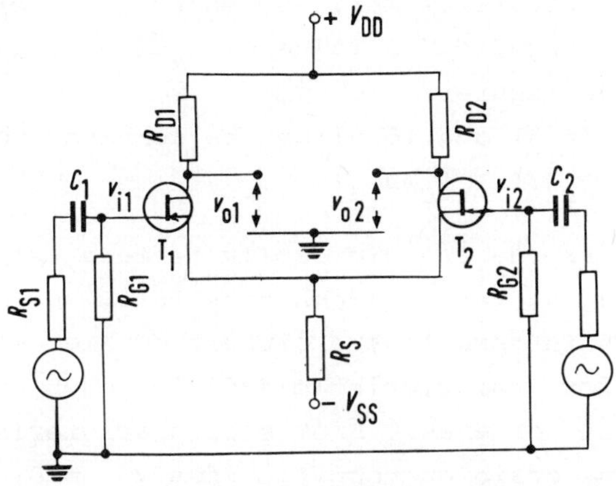

Fig. 6.3 A differential amplifier utilizing n-channel fets.

where A_1 is the voltage gain from input 1 to the output with input 2 connected to the common line and A_2 is defined similarly. Ideally $A_1 = A_2$ but this exact equality cannot be achieved on a practical differential amplifier, although it can be closely approached by the use of the monolithic dual fet construction.

Now

$$v_{i1} = v_i + \tfrac{1}{2} v_d \quad \text{and} \quad v_{i2} = v_C - \tfrac{1}{2} v_d$$

Equation (6.1) therefore becomes

$$v_o = A_d v_d + A_C v_C \tag{6.3}$$

where A_d is the voltage gain for the difference signal and is defined by

$$A_d = \tfrac{1}{2} (A_1 - A_2) \tag{6.4}$$

and A_C is the voltage gain for the difference signal and is defined by

$$A_c = A_1 + A_2 \qquad (6.5)$$

A_d should be very large and A_c very small, ideally zero. The ratio A_d/A_c is ρ, the common-mode rejection ratio (cmrr). Hence

$$\rho = A_d/A_c \qquad (6.6)$$

ρ serves as a figure of merit by which a differential amplifier is assessed. A value of ρ of 5000 is practicable as a high figure.

Equations (6.3) and (6.6) can be combined to give

$$v_0 = A_d v_d \, (1 + v_c/\rho v_d) \qquad (6.7)$$

6.1.1 *To determine the similarity between two fets fabricated on the same silicon substrate.* Using each fet of the 2N3955 in turn in the circuit of Fig. 6.4, plot the static transfer characteristic and plot each characteristic in the same set of axes. From each characteristic deduce values for the drain current I_{DSS} for $V_{GS} = 0$; the pinch-off voltage V_P; and g_m, the mutual conductance over the linear portion of the characteristic (say, at $V_{GS} = -0.3V$). Use these values of V_P and I_{DSS} to check the parabolic nature of the transfer characteristic given by the equation

$$I_D = I_{DSS} \, (1 - V_{GS}/V_P)^2$$

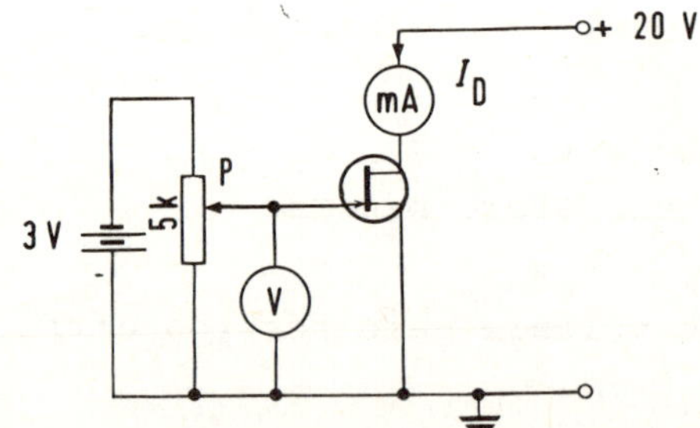

Fig. 6.4 Determination of the transfer characteristic of each fet in a pair fabricated on the same silicon substrate (eg 2N 3955). V is a high resistance voltmeter which measures V_{GS}.

Compare the experimental and calculated values of I_D for three values of V_{GS}.

6.1.2 *The phase relationships in the differential amplifier.* Set up the circuit of Fig. 6.5 in which the two resistors R_D are a matched pair and all the resistances should be within 1 per cent of the stated values. Set the voltage V_{SS} at approximately -16.5 V and adjust it slightly until the voltage at the point P is at +0.3 V with respect to earth. The operating point of each fet has thus been chosen to be at V_{GS} = 0.3 V.* Measure the voltage drop across R_S and establish that the current through it is $2I_D$. With v_{i2} and v_{i1} = 50 Vrms note by the use of a double-beam cro that v_{01} is 180° out of phase with v_{i1} whereas v_{02} is in phase with v_{i1}. Note also there is an amplified signal at the drain of T_2 even though the input signal to T_2 is zero.

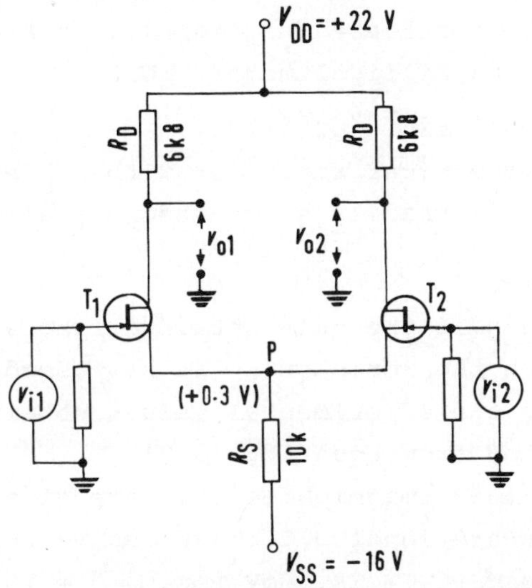

Fig. 6.5 A basic fet differential amplifier stage

*This is a suitable value for the device used here. Students should select a suitable value from the mutual characteristics.

6.1.3 *The paraphase amplifier.* With the gate of the transistor T_2 earthed (ie $v_{i2} = 0$) and $v_{i1} = 50$ mV rms, display and record the waveforms of the two output signals v_{01} and v_{02} on the double beam oscilloscope: note that the two output signal voltages are of equal amplitude but 180° out of phase.

Such an amplifier stage, providing two output signals from a single input, is called a paraphase amplifier. It is frequently used to provide two voltage signals, one to each power transistor, in the output stage of an audio frequency amplifier.

6.1.4 *The differential voltage gain A_d.* Set $v_{i2} = 0$ (ie earth the gate of the transistor T_2) and make $v_{i1} = 5$ mV rms (measured with a suitable ac millivoltmeter) at a frequency of 1 kHz. Then $v_d = v_{i1}$ and $v_o = A_d v_d$ because A_c and v_o of equation (6.3) are both small and their product can be considered negligible. Measure v_{o1} and v_{o2} with the same ac millivoltmeter. Now

$$A_d \simeq -v_{o1}/v_{i1} = v_{o2}/v_{i1} \tag{6.8}$$

Assuming a symmetrical amplifier with R_s large compared with $1/g_m$, it is readily shown theoretically that

$$A_d = g_m R_D/2 \tag{6.9}$$

which is just half the gain obtained from a single CS stage in which the resistor R_s is by passed with a capacitor. Compare the experimental values obtained for A_d with that calculated from equation (6.9).

Typical result (magnitudes only are given, phase relationships are ignored). Measured values: $v_{i1} = 5$ mV rms, $v_{o1} = 19.5$ mV rms, $v_{o2} = 18$ mV rms.

(i) $A_d = 19.5/5 = 3.9$;

(ii) $A_d = 18/5 = 3.6$

Calculated value: $A_d = g_m R_d/2 = 1.1 \times 6.8/2 = 3.7$

6.1.5 *The common-mode voltage gain* A_c. If the differential amplifier shown in Fig. 6.5 were ideal, v_{o1} and v_{o2} would be zero when $v_{i1} = v_{i2}$. In practice, equation (6.3) applies:

$$v_o = A_d v_d + A_c v_c$$

With $v_{i1} = v_{i2}$, the difference voltage $v_d = 0$ and the output voltage would be determined by the common-mode signal

$$v_c = \tfrac{1}{2}\,(v_{i1} + v_{i2})$$

and the common-mode voltage gain

$$A_c = -\, v_o/v_c\,.$$

To determine A_c by experiment, apply the same sinusoidal signal of 50 mV rms and frequency 1 kHz (from a single signal generator) to each gate of the amplifier of Fig. 6.5. Display and record the output voltage waveforms v_{o1} and v_{o2} by means of a double beam oscilloscope. Note that these two output signals are equal in amplitude and are in phase whereas both output voltages are 180° out of phase with the respective input signals. Measure v_{i1}, v_{o1} and v_{o2} with a suitable ac millivoltmeter. Increase the magnitudes of the input signals (keeping them equal) in 50 mV steps up to 400 mV and measure v_{o1} and v_{o2} at each setting. Plot a graph (typically as in Fig. 6.6) of $v_o = v_{o1} = v_{o2}$ against the common-mode signal voltage v_c. Calculate from the graph a value for A_c where

$$|A_c| = v_c/v_o$$

and compare with the result given for an ideal differential amplifier (Fig. 6.5) of

$$|A_c| = R_D/2R_s = \frac{-6.8\times10^3}{2\times10^4} = 0.34$$

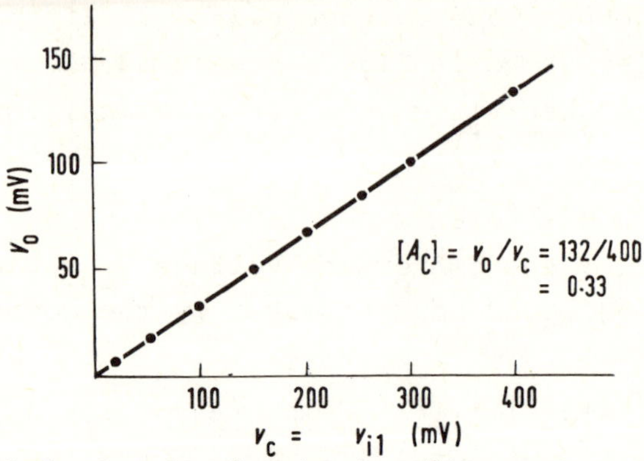

Fig. 6.6 Graph of output voltage v_o against the common-mode signal voltage v_c for the amplifier of Fig. 6.5

6.1.6 *The common-mode rejection ratio (cmrr).* A_d should be as large as possible and A_c as small as possible if the differential amplifier is to approach the ideal. The ratio A_d/A_c - called the common-mode rejection ratio - is used to assess the quality of a differential amplifier where clearly it should be as large as possible. The cmrr for the amplifier shown in Fig. 6.5 can be calculated from results obtained in previous experiments: from experiment 6.1.4 typically $A_d = 3.75$; from experiment 6.1.5, typically $A_c = 0.33$. Hence

$$\text{cmrr} = 3.75/0.33 = 11.4$$

which compares with the theoretical value given by

$$\text{cmrr} = A_d/A_c = (g_m R_d/2)/(R_d/2R_s) = g_m R_s$$
$$= 1.1 \times 10 = 11.$$

In a further experiment, set up the circuit of Fig. 6.5 with R_s = 15 kΩ and increase the voltage V_{SS} to - 26 V approximately in order to maintain the correct bias conditions (V_p = +0.3 V). Repeat the procedure of the experiment in Sections 6.1.4 and 6.1.5 to find A_d and A_c for this circuit and hence the cmrr.

It will be readily shown by these further experiments that - as would be expected from simple theory - the cmrr is increased by increasing R_S. However, this requires considerably increased values of V_{SS} in order to maintain the correct bias conditions. It becomes unrealistic to attempt to increase further the cmrr by simply increasing the magnitude of R_S.

6.1.7 *Use of a constant current source to replace* R_S. This enables a differential amplifier with a very large value of cmrr to be made. In Fig. 6.7 a differential fet amplifier is shown with a bipolar transistor (BC 109) as a constant current source: the dynamic resistance of this constant current source can be of the order of several megohms yet retaining realistic values of V_{SS}.

In an experiment, the circuit of Fig. 6.7 is used with R_E = 1 kΩ. The value of V_{SS} is adjusted with $v_{i1} = v_{i2} = 0$ until the steady pd across each drain resistor is 6.12 V (0.9 mA × 6.8 kΩ). Under these conditions, the constant current is 1.8 mA ($2I_D$) and V_{GS} for each fet is -0.3 V and I_D = 0.9 mA. The diodes D_1 and D_2 are used to provide temperature compensation for the emitter-base junction of the bipolar transistor BC 109.

With v_{i1} = 5 mV and v_{i2} = 0, determine the differential voltage gain A_d for this amplifier (Fig. 6.7) and with v_{i1} and v_{i2} = 300 mV, determine the common mode voltage gain A_c. Use these values to calculate

$$\text{cmrr} = A_d/A_c$$

Typical results

(i) R_E = 1 kΩ, voltage across R_D = 6.12 V with V_{SS} = - 5.5 V. Hence I_D = 0.9 mA and V_{GS} = 0.3 V.

With v_{i1} = 5 mV and v_{i2} = 0, A_d = 3.7

With $v_{i1} = v_{i2}$ = 300 mV, $v_{o1} = v_{o2}$ = 0.66 mV. Thus

$$A_c = 0.66/300 = 2.2 \times 10^{-3},$$

therefore

$$\text{cmrr} = A_d/A_c = 3.7/(2.2 \times 10^{-3}) = 1680$$

(ii) For R_E = 2 kΩ, cmrr = 3460

(iii) For R_E = 3 kΩ, cmrr = 5110

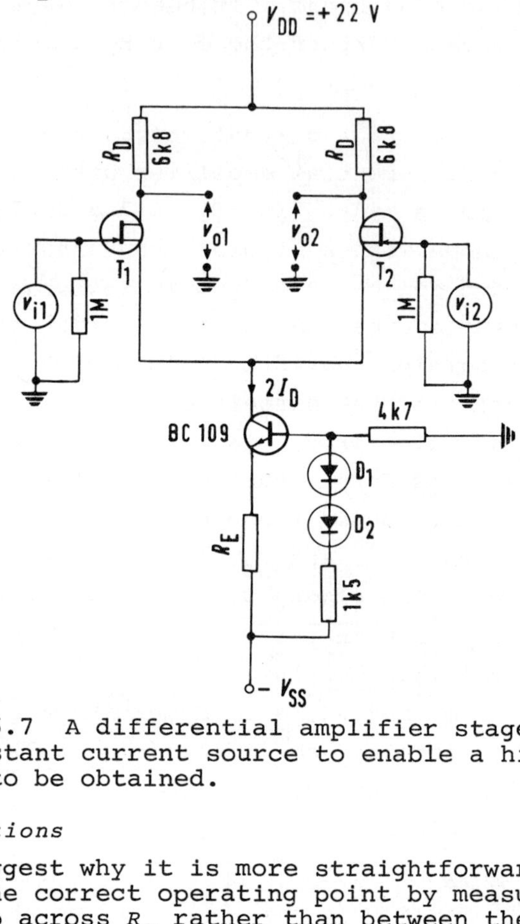

Fig. 6.7 A differential amplifier stage with a constant current source to enable a high cmrr to be obtained.

6.1.8 *Questions*

(i) Suggest why it is more straightforward to establish the correct operating point by measuring the voltage drop across R_D rather than between the joined sources of the FETs T_1 and T_2 (Fig. 6.7) and earth when the constant current source is connected.

(ii) For a cmrr of this stage of 1680, estimate the effective resistance of the constant current source, utilizing the results of Experiment 6.1.6.

(iii) If the resistance calculated in question (ii) were used as a passive component, what value of V_{SS} would be required to establish the correct operating point of the fet?

(iv) In Fig. 6.7, assuming that the base current of the bipolar transistor BC 109 is negligible and that each junction voltage is 0.7 V, calculate the constant current through the BC 109 for R_E = 1 kΩ and V_{SS} = -5.5 V.

6.2 *Operational amplifiers*

Three forms of operational amplifier are available

(a) Discrete modular: a unit in which discrete components are selected, mounted on a small printed circuit board and the whole unit encapsulated; the dimensions of these units are about 25 mm × 25 mm × 12 mm.

(b) Hybrid: a somewhat smaller unit made up partly of discrete components with the remainder an integrated circuit.

(c) Integrated circuit (ic) form: the complete amplifier is constructed on a single silicon slice (a monolithic integrated circuit) mounted in a TO99 can (an 8-pin reduced height TO5 can) or a dual in-line package (ie dil or dip).

The discrete modular types - which are the most expensive - offer slightly better characteristics. It is by no means essential for a user of operational amplifiers to be familiar with the intricate internal circuit details, but the arrangement of the external terminals and their functions and the terms used to specify the amplifier performance must be understood to enable intelligent selection for a particular application to be made of an operational amplifier.

The experiments to be described here allow some of the features of operational amplifiers to be examined and a number of simple applications to be introduced.

A discrete modular unit chosen is the E78, a general purpose differential amplifier with a dual fet input. A differential input amplifier enables greater flexibility in the choice of the feedback configuration than a single-ended input amplifier. The output voltage of an ideal differential input amplifier depends only on the difference between the voltages applied to the two input terminals. Any common voltage with respect to the common line that appears at the input terminals is called the common-mode input voltage; in the ideal amplifier, this will have no effect on the output voltage. The ideal amplifier is also assumed to have

(a) infinite gain so that the performance will be entirely dependent on the input and feedback networks. The E 78 has a minimum dc open-loop voltage gain (ie gain without feedback) of 10^5;

(b) infinite input resistance so as to ensure that no current flows into the amplifier input terminals. For the E 78 the input impedance is 10^{12} Ω both differential and common-mode;

(c) infinite bandwidth so as to ensure a response to dc signals, zero response time and zero phase change with frequency. The E 78 provides amplification without distortion up to 100kHz where this gain will decrease as the frequency increases; the transition frequency (at which the gain drops to unity) is 3 MHz.

The input terminals of the operational amplifier are labelled - and +. The - terminal is the input to the phase inverting amplifier; the + terminal is the input to the non-inverting amplifier.

The two basic configurations with feedback applied to a differential amplifier are an inverting one (Fig. 6.8a) and a non-inverting one (Fig. 6.8b). In each case the signal is fed back via resistor R_2 to the point X - the phase inverting input terminal, sometimes called the amplifier summing point. (Note that point X becomes point B in Fig. 6.8b).

If the input voltage V_i tends to drive the potential positive with respect to earth at point X, the right-hand end of R_2 assumes a negative potential with respect to earth due to the inverting nature of an amplifier. A current I_f will flow in R_2 to reduce the positive excursion of X to a very low value. With an amplifier voltage gain of 10^5 or more, an excursion of a few microvolts at X is sufficient to cause the output voltage to swing through its entire range - perhaps 10 V on either side of earth. The self-adjusting nature of the negative feedback system has the effect of holding the potential at X to a very low value, so $V_e \to 0$ no matter what values are chosen for R_1 and R_2. The point X is thus often called a virtual earth.

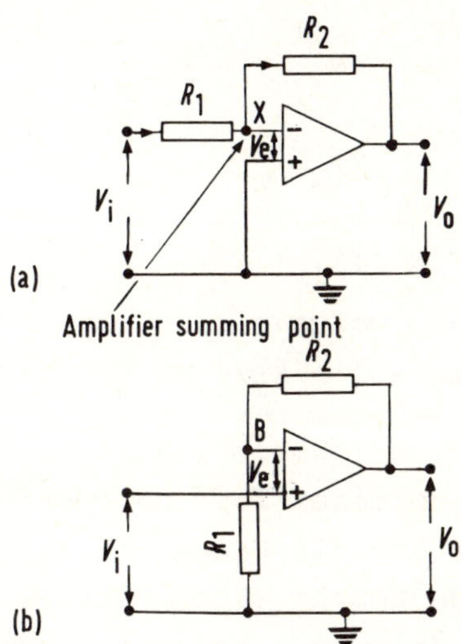

Fig. 6.8 The two basic ways by which feedback may be applied to a differential amplifier: (a) in the inverting configuration; (b) in the non-inverting configuration.

6.2.1 *The closed-loop gain of the inverting configuration.* Assume that V_e is zero, ie an ideal differential amplifier is concerned. Then, in Fig. 6.8a

$$I_1 = V_i/R_1 \text{ and } I_f = -V_o/R_2$$

where I_1 is the current through R_1 and I_f is the current through R_2. But since the current flow into the amplifier is zero,

$$I_i = I_f$$

Therefore

$$-V_o/R_2 = V_i/R_1$$

and the closed-loop voltage gain

$$A_{VCL} = V_o/V_i = -R_2/R_1. \quad (6.10)$$

6.2.2 *The closed-loop gain of the non-inverting configuration.* Because the current flow into either input of the amplifier is zero, the voltage V_B at the inverting input terminal (Fig. 6.8b) is given by

$$V_B = V_o R_1/(R_1 + R_2)$$

As

$$V_e = 0$$

therefore

$$V_i = V_B = V_o R_1/(R_1 + R_2)$$

and the closed loop voltage gain is

$$A_{VCL} = V_o/V_1 = 1 + R_2/R_1. \quad (6.11)$$

6.2.3 *The operational amplifier as a sign inverter.* The necessary supply voltages for the operational amplifier module (type E78) are +15 V and -15 V from a suitably stabilized voltage supply (such voltage supplies are never shown in circuit diagrams, but are assumed, as also is the input offset voltage control in the operational amplifier). Batteries can be used; indeed, this operational amplifier will work satisfactorily with voltages down to +10 V and -10 V, although with decreased gain. It is important that the zero point (between +15 and -15 V) of the power supply be connected to the common line (earth). Connect up the E78 (with its power supply connected to the appropriate pins) in the circuit of Fig. 6.8a with, initially, $R_1 = R_2 = 10$ kΩ (obtained from appropriate resistance boxes). Connect together the input terminals and vary the zero adjustment potentiometer (maximum value 1 kΩ) which is the input offset voltage control, until the output voltage is zero. As this input offset voltage is slightly temperature sensitive, it should be checked at intervals throughout the experiments.

Set up the circuit of Fig. 6.9 with $R_1 = R_2 = 100$ kΩ. With the input potentiometer R apply steady voltages within the range 0 to 0 ± 10 V and for each known input voltage V_i record the output voltage V_o. Plot V_o against V_i. In this form, with $R_1 = R_2$, the operational amplifier has a gain of unity (equation 6.10) and acts as a sign inverter.

6.2.4 *The gain control in the inverting configuration.* Using the circuit of Fig. 6.9 vary the gain of the amplifier by changing the ratio R_2/R_1 and, at each setting, measure both the input voltage V_i and the output voltage V_o. Tabulate the results and compare the measured voltage gain ($A_{VCL} = V_o/V_i$) with that calculated from $-R_2/R_1$ (equation 6.10). It is important here to take into account the accuracy with which V_i, V_o, R_1 or R_2 have been measured.

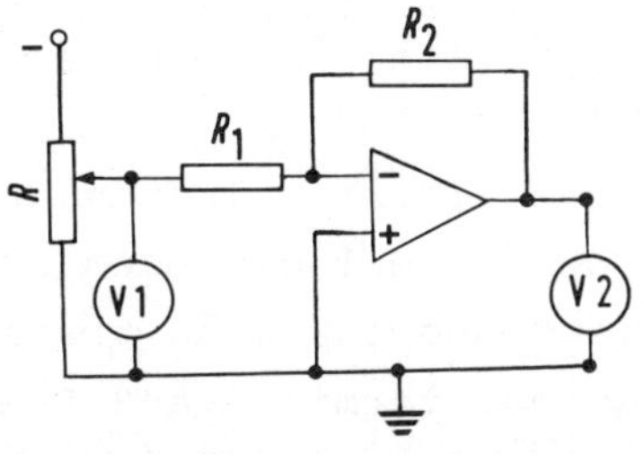

Fig. 6.9 Experiments on the inverting configuration (see also Fig. 6.8a). Voltmeter V1 measures V_i; voltmeter V2 measures V_o.

Set $R_2 = 1$ MΩ and $R_1 = 10$ kΩ so that the voltage gain calculated from equation (6.10) is $A_{VCL} = -10^6/10^4 = -100$. Use the potentiometer R and an appropriate voltage supply to apply small steady voltages (both positive and negative) and measure V_i and V_o for each setting of V_i.

Plot a graph of the measured values of V_o against the calculated values of V_o from $V_i\ A_{VCL}$. A typical curve is as in Fig. 6.10. Note that, with a +15 V,0, -15 V power supply, the output voltage is limited to ± 14 V approximately. If linear behaviour is to be assumed, the amplifier input should always be maintained at a level which will not overload the amplifier.

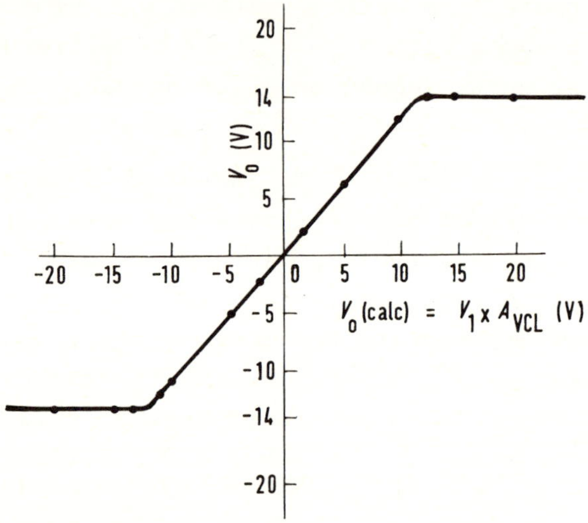

Fig. 6.10 Typical operational amplifier characteristic.

6.2.5 *Further investigation.* In the two experiments described so far the current demanded from the amplifier is virtually zero. Connect a load resistance R_L of value 2 kΩ between the output terminal and the common line. Repeat the experiments described while arranging for the output current to be maintained at 10 mA. Tabulate these results. Do they differ from those obtained previously?

6.2.6 *The frequency response of the amplifier in the inverting configuration.* Replace the dc potentiometer of Fig. 6.9 with a signal generator providing a signal voltage of sinusoidal waveform of which both the frequency and the amplitude can be varied. Set the amplifier gain to unity with $R_1 = R_2 = 100$ kΩ (using carbon resistors for R_1 and R_2), connect a load resistance $R_L = 2$ kΩ between the output terminal and the common line and with $V_i = 2$V rms (as obtained from the signal generator and measured by an ac millivoltmeter) measure with an ac millivoltmeter the output voltage V_o over the frequency range 10 Hz to 1 MHz. Observe and record any distortion or phase shift introduced by the amplifier. Plot a graph of the voltage gain (V_o/V_i) against the frequency. Repeat with a gain (decided by R_2/R_1) of 10 and then 100. In each case, adjust the amplitude of the input signal so that it does not overload the amplifier. Typical graphs obtained are as in Fig. 6.11.

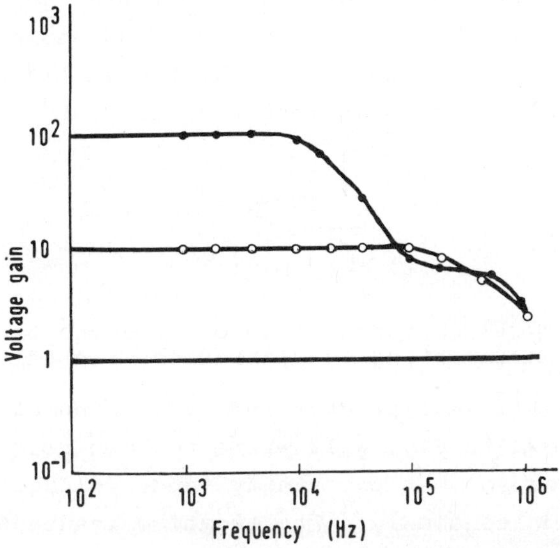

Fig. 6.11 Graphs of voltage against frequency (frequency response curves) for an operational amplifier (E78 in the inverting mode).

6.2.7 *The closed-loop voltage gain in the non-inverting configuration.* Using the non-inverting circuit of Fig. 6.8b set $R_1 = R_2 = 100$ kΩ and for steady input voltages between 0 and 5 V, measure V_o; use the results to verify equation (6.11).

Gain control in the non-inverting configuration. With the circuit of Fig. 6.8b set R_1 = 1 kΩ and R_2 = 1 MΩ so that the calculated closed loop voltage gain $A_{VCL} = 10^3$. Apply small steady voltages at the input; measure V_i and V_o; plot the measured value of V_o against the calculated value $V_o = V_i A_{VCL}$.

6.2.8 *Frequency response in the non-inverting configuration.* Examine the frequency response with gain unity making use of the circuit of Fig. 6.12, in which $R_2 = 0$ and $R_1 = \infty$, so that the gain is unity (as calculated from equation 6.11). With a load resistance R_L = 2 kΩ, apply from a suitable signal generator an input signal of 2 V (peak to peak) and measure V_o. Determine the voltage gain over the frequency range 10 Hz to 1 MHz; note any distortion or phase shift introduced by the amplifier; plot a graph of voltage gain against frequency.

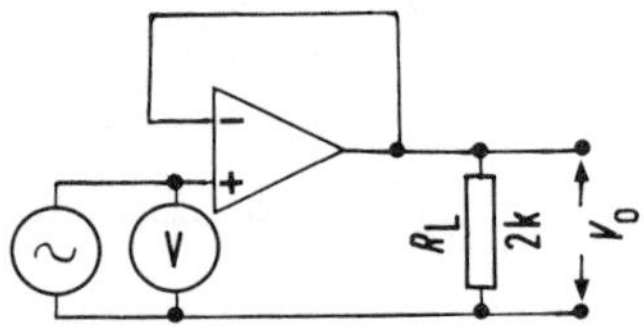

Fig. 6.12 Determination of frequency response of non-inverting configuration with unity gain.

6.2.9 *Simultaneous product and sum.* Connect up the circuit shown in Fig. 6.13 where the two potentiometers are arranged so that two steady input voltages may be applied simultaneously. The negative feedback condition ensures that the output voltage in this configuration is the inverted sum of the two applied voltages. Also, either or both of the input voltages may be multiplied by a constant by varying the ratio of the individual series resistors to the common feedback resistor. In this way any number of voltages may be added. In any one operation it is possible to multiply each input voltage by a constant and to add the new voltages together. Labelling one input as *x* and one as *y* use this circuit (Fig. 6.13) to investigate the solution of the equation

$$5x + 10y = -z$$

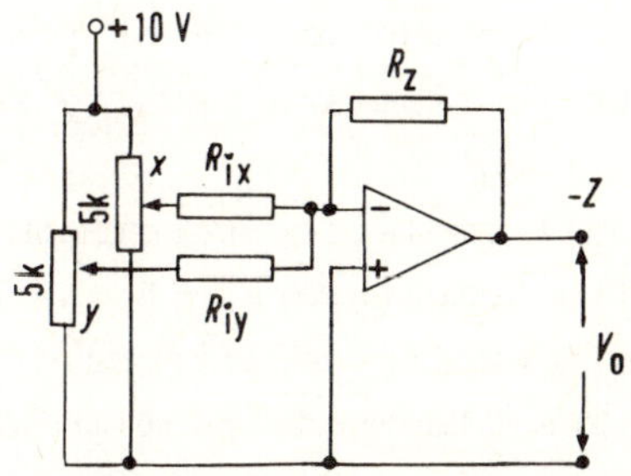

Fig. 6.13 Simultaneous product and sum.

Note: keep the common feedback resistor R_2 at a value of 1 MΩ .

6.2.10 *Comment.* The chief difference between the inverting and the non-inverting configuration lies in the effective input resistance which the circuits present to the signal source. In the ideal inverter, feedback prevents the voltage at the summing point from changing - the point acts as a virtual earth. Any current supplied at this point flows through the feedback resistor R_2. The input current I_i is therefore determined by R_1 so that the input resistance presented to the signal source is R_1. In the ideal non-inverting circuit no current is taken from the signal source so that the input resistance is virtually infinite.

6.2.11 *Integrated circuit operational amplifiers.* The ic operational amplifiers in common use are monolithic, ie fabricated on a single crystal silicon slice and where epitaxial, planar and diffusion processes are used in the manufacture. The two encapsulations generally available are

(i) TO99 metal can - an 8-pin reduced height TO5, often called a T package.

(ii) TO116 dual-in-line (dil) encapsulation where the 8-pin version is called a V package and the 14-pin one is known as an A package.

The body of the dil package may be of plastic or ceramic, although the latter is rare and usually reserved for military use.

The ic operational amplifier recommended for experiment is the 741. As with transistors, it has been agreed internationally to identify amplifiers of the same type by the same number. The electrode configuration of the 741 (a V package) is shown in Fig. 6.14a while the offset compensation circuit is seen in Fig. 6.14b. In Fig. 6.14a the device is viewed from above and one end if identified by a recess in the plastic body. This recess should always be aligned with a similar recess or chamfer on the socket to ensure correct fitting. Viewed from above, with the recess on the left-hand side, the pin numbers are always arranged as in Fig. 6.14a with electrode 1 in the bottom left-hand corner and subsequent numbers proceeding anti-clockwise. On some devices no recess exists on the package but a dot is located against pin 1. Relevant equations for an operational amplifier are (6.10) and (6.11) for the inverting and non-inverting configurations respectively.

Table 6.1 gives some of the characteristics of the 741.

Table 6.1 Characteristics of the 741.

Characteristic	Ideal	741
Large signal open loop gain, A_{VOL}	infinite	100 dB or 10^5
Input resistance, R_i	infinite	2MΩ
Common-mode rejection ratio	infinite	30 000
Differential input offset current, I_{io}	zero	20 nA
Differential input offset voltage, V_{io}	zero	1 mV

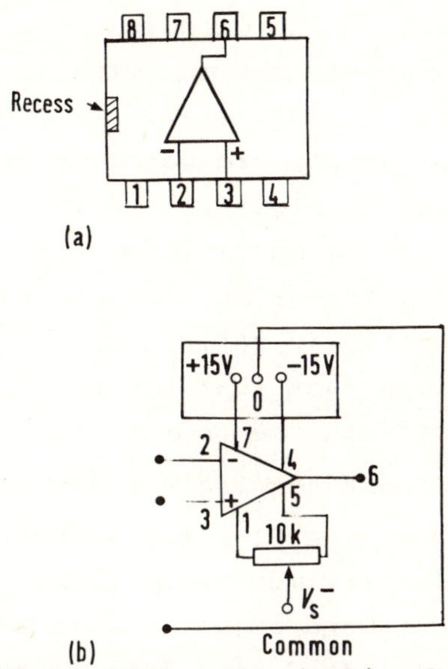

Fig. 6.14 (a) Plan view (top) of ic operational amplifier 741; (b) the offset compensation circuit for the 741.

The 741 has internal frequency compensation and its output current is limited to 10 mA so that the device is protected if its output terminals are short-circuited. This model of operational amplifier performs well in a number of applications; it is, however, a low-cost device having a performance much inferior to many of the operational amplifiers now available.

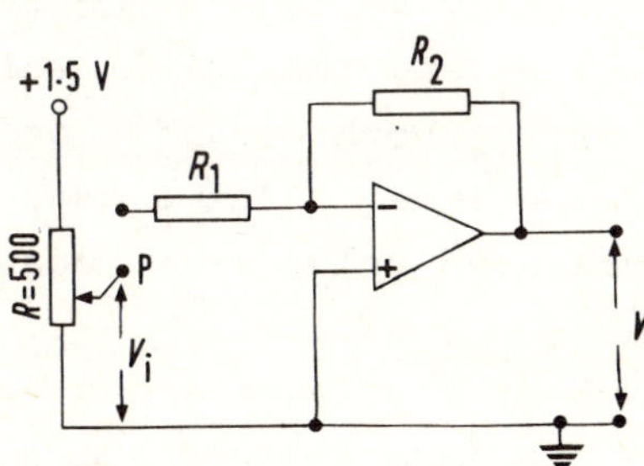

Fig. 6.15 The operational amplifier (741) in the inverting configuration.

6.2.12 *Transfer characteristic of the amplifier in the inverting configuration.* The circuit (Fig. 6.15) to be used for this experiment is set up with $R_1 = R_2 = 10\ k\Omega$. The appropriate supply voltages (+15V to 0 to -15V) from a suitable stabilized supply are connected (these voltage supplies are not shown in Fig. 6.15 or in subsequent circuit diagrams involving operational amplifiers). Two 12 V or two 9 V dry batteries may be used to provide these supply voltages resulting in satisfactory performance but lower gain. It is important that the zero point (0) of the supply be connected to the common line.

(a) Connect the input terminals together (Fig. 6.15) with the input voltage $V_i = 0$. Connect the zero adjustment potentiometer (of maximum value 10 kΩ; see Fig. 6.14b) into the circuit and adjust its value until the output voltage (as indicated by a dvm) is zero. This input offset voltage is slightly temperature sensitive and so should be checked at intervals during the experiment.

Set $R_2 = 100\ k\Omega$ and $R_1 = 10\ k\Omega$ so that the closed loop voltage gain $A_{VCL} = -R_2/R_1 = -10^5/10^4 = -10$. Decade resistance boxes may be used for R_1 and R_2 but if separate resistors are used the resistance of each should be known to within 1 per cent. Connect the point A to the point P and apply a small steady voltage V_i to the input of the amplifier. Measure V_i and the output voltage V_o for a number of values of V_i, both positive and negative with respect to the common line. Plot a graph (Fig. 6.16 is typical) of the measured value of V_o against the calculated value, ie of $V_{o(obs)}$ against $V_{o(calc)}$ given by $V_i\ A_{VCL}$. Note that the graph of Fig. 6.16 shows that saturation does not occur at the same voltage for a positive input as for a negative input, ie the transfer characteristic of the amplifier is not perfectly symmetrical.

If linear behaviour is to be assumed, the input signal

must always be at a level which will not cause the amplifier to saturate.

(b) The current demanded from the amplifier was virtually zero during experiment (a) just described. Connect a decade resistance box set initially at 2 kΩ as a load resistance R_L in series with a milliammeter (0 - 10 mA) between the output terminal of the operational amplifier and the common line. Repeat some of the readings of experiment (a) and vary R_L to arrange for the output current through R_L to be maintained at 10 mA. Compare the results with those obtained previosuly.

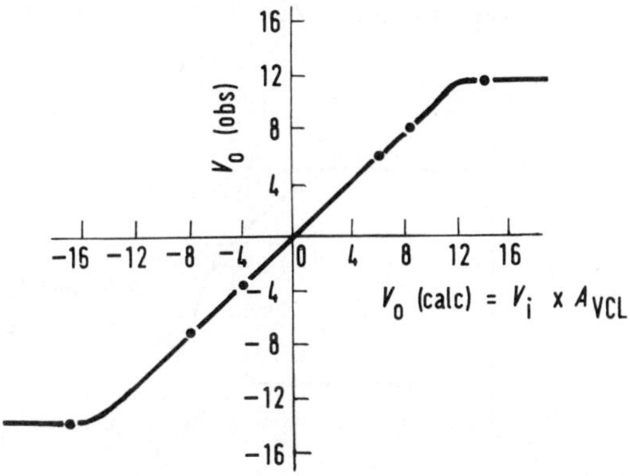

Fig. 6.16 A typical transfer characteristic for a 741.

(c) Repeat experiment (a) with a resistor R_3 connected between the non-inverting terminal and the common line where $R_3 = R_1R_2/(R_1 + R_2)$. This is to equalize the dc resistance at the input terminals and hence reduce the error created by the input offset voltage and current.

With this resistor R_3 in position, is the transfer characteristic obtained in the experiment more symmetrical?

Note. Any of the experiments based on the E78 - discrete modular type of operation amplifier - may be repeated with the ic operational amplifier 741.

6.2.13 *Frequency response of the amplifier in the inverting configuration.* An amplifier is said to have a first order frequency response if its voltage gain A_f at frequency f is related to the gain A_L at low or zero frequency by the equation

$$A_f = \frac{A_L}{1 + jf/f_1} \quad (6.12)$$

where f_1 is termed the first corner frequency or the 3 dB frequency. This equation (6.12) can be written in the form

$$A_f = A_L \ \exp(-j\phi)/\sqrt{|1 + (f/f_1)^2|} \quad (6.13)$$

where $\phi = \tan^{-1}(f/f_1)$

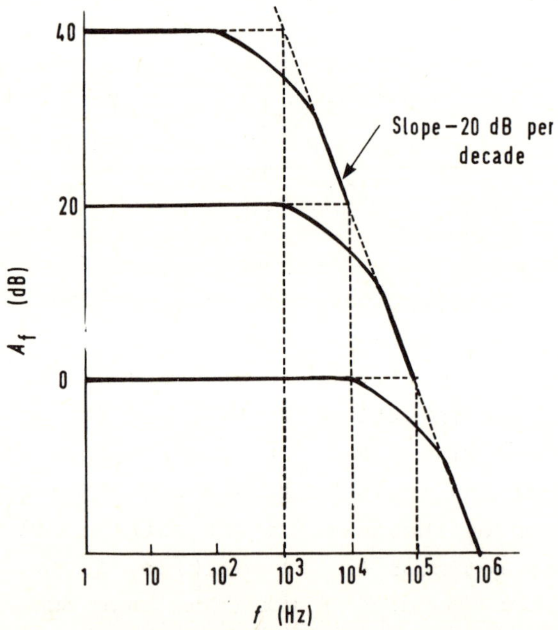

Fig. 6.17 Idealized frequency response curves for an operational amplifier

In Fig. 6.17 the magnitude $|A_f|$ is plotted against the frequency f. At low frequency for which $f < f_1$, the horizontal line is the low frequency asymptote. At $f = f_1$, the gain is 3 dB down on the low frequency value and at higher frequency the gain drops at 20 dB per decade. Because the high frequency asymptote has a negative slope of 20 dB per decade, the gain of the amplifier is said to 'roll-off' at 20 dB per decade.

Utilizing the circuit of Fig. 6.15, replace the potentiometer by a signal generator providing an input voltage V_i of sinusoidal waveform of which both the frequency and the amplitude can be varied. By making $R_1 = R_2 = 100$ kΩ, set the gain of the operational amplifier to unity. Connect a load resistance R_L of 2 kΩ and with $V_i = 2$ V peak-to-peak, measure with a suitable ac millivoltmeter the output voltage V_o over the frequency range 10 Hz to 1 MHz.

In such frequency response measurements, carbon resistors are preferable to decade resistance boxes for R_1 and R_2. Observe and record by means of a cro any distortion or phase shift introduced by this amplifier. Plot a graph of the voltage gain V_o/V_i against the frequency. Repeat with ratios of $R_2/R_1 = 10$ and then 100. In each case, adjust the amplitude of the input signal voltage so that the output voltage does not exceed 2 V peak-to-peak Represent the frequency response curves obtained by straight line segments and establish that the high frequency asymptote has a slope of 20 dB per decade.

6.2.14 *Slew-rate limiting of the amplifier.* There is a limit to the rate at which the output voltage of an amplifier can change, set by the presence of internal capacitance and the movement of charge carriers by diffusion in the base region of bipolar transistors. The maximum possible rate of change of the output voltage is called the slew rate S.

For a sinusoidal input signal represented by

$$V = V_P \sin\omega t$$

$$\frac{dV}{dt} = V_P \, \omega \cos \omega t$$

and this has a maximum value when $\omega t = 0$, so $t = 0$. Therefore

$$\frac{dV}{dt}_{max} = 2 \pi f V_P$$

where f is the frequency $\omega/2\pi$

For an amplifier to operate with a sinusoidal input signal without distortion, S must considerably exceed $2\pi f V_P$. So slew rate limiting or distortion created by an inadequate slew rate depends on both the frequency and the amplitude of the output signal.

Set up the operational amplifier in an inverting configuration (Fig. 6.15) with $R_1 = R_2 = 100$ kΩ (gain: unity) Apply an input signal of sinusoidal waveform of magnitude 0.5 V rms from a single generator. With a load resistance $R_L = 2$ kΩ, measure the output voltage using a suitable ac millivoltmeter. Determine the highest frequency of the input signal at which the gain is seen to be unity ($V_o = V_i$). Increase the input signal to 1 V rms and again find the highest input signal frequency at which $V_o = V_i$. The ac millivoltmeter recording the output voltage is being used to indicate the onset of the distortion created by inadequate slew rate. The input and output waveforms can be examined with a cathode ray oscilloscope but it is very difficult to detect a small amount of distortion. Continue this procedure, increasing the input voltage in steps of 0.5 V up to, say, 4 V rms.

In each case the maximum frequency at which unity gain is maintained is lower for larger amplitude signals. Assuming that the bandwidth is not the limitation, an inadequate slew rate must be responsible for the distortion. Fig. 6.18 shows the reciprocal of the peak output voltage plotted against the highest frequency at which unity gain was maintained for a 741 operational amplifier.

From this graph, the slew rate of the amplifier is estimated to be

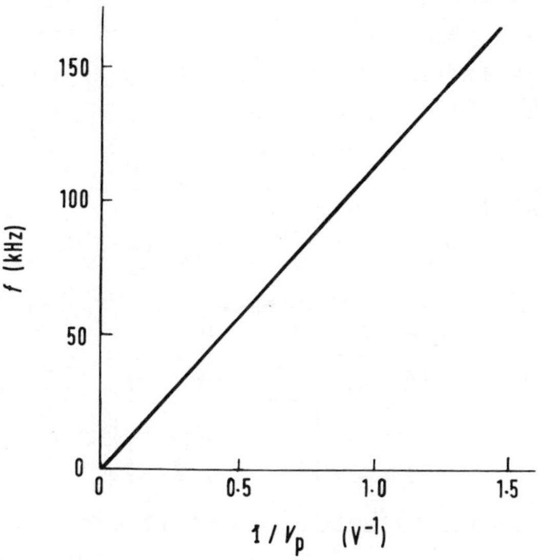

Fig. 6.18 Slew rate limiting of a 741

$$S = 2\pi f V_P$$

$$= \frac{2\pi \times 150 \times 10^3}{1.33} \text{ Vs}^{-1} = 0.71 \text{ V}\mu\text{s}^{-1}$$

There are other methods of measuring slew rate but this experiment emphasises that distortion created by an inadquate slew rate depends on both the frequency and the amplitude of the output voltage. Generally speaking, an amplifier may be designed for either a large output voltage or a high frequency performance, but not both.

6.2.15 *A logarithmic amplifier.* The use of a silicon diode as the feedback element enables an operational amplifier to be created in which the output voltage is proportional to the logarithm of the input voltage (Fig. 6.19). This transpires because the current-voltage relationship for a p-n junction (as in a silicon diode) is

$$I = I_s \; |\exp(eV/kT) - 1|$$

where I_s is the reverse saturation current, V is the voltage across the junction, whilst k, e and T have their usual meanings. At T = 300 K, kT/e = 26 mV approximately; for voltages V in excess of about 100 mV, therefore, the equation reduces to approximately

$$I = I_s \exp (eV/kT)$$

ie $$\log (I/I_s) = eV/kT$$

and $$V = 2.3 (kT/e) \log_{10} (I/I_s).$$

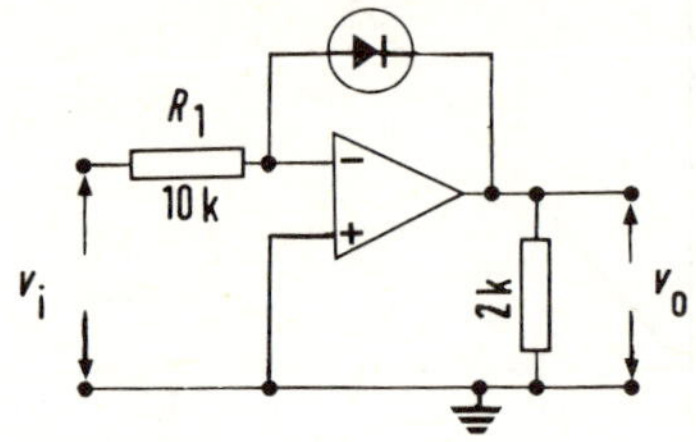

Fig. 6.19 A logarithmic amplifier

Considering the operational amplifier of Fig. 6.19 to be 'ideal', the voltage V across the diode is equal to $-V_o$ because the inverting input terminal is a 'virtual' earth. Therefore

$$V_o = 2.3(kT/e) \log_{10}(I/I_s)$$

but $$I = V_i/R_1$$

and $$V_o =-2.3(kT/e) \log_{10}(V_i/R_1 I_s).$$

Thus, using the circuit of Fig. 6.19 apply a stable steady input voltage V_i within the range 1 mV to 10 V and measure both the input and the output voltages with a digital voltmeter. Plot a graph of V_i (on a logarithmic scale) against V_o (linear scale) to give typically a graph as in Fig. 6.20.

The factor $2.3(kT/e)$ has a value of approximately 60 mV per current decade. Confirm this fact by calculation and from the graph. It is often convenient to have the output voltage increase by one volt per current decade and this can readily be achieved by using the circuit of Fig. 6.21,

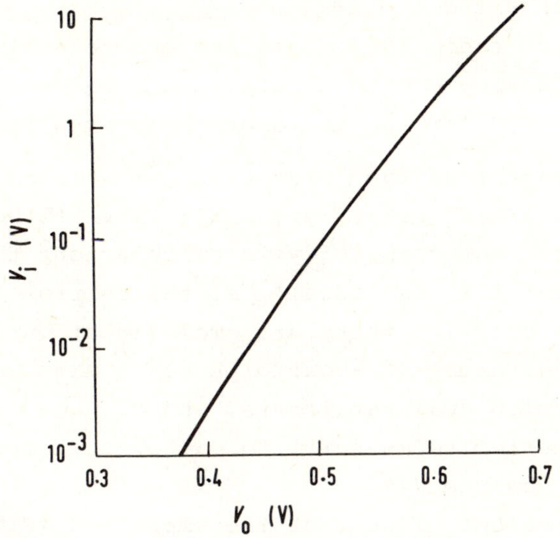

Fig. 6.20 Typical characteristic of the logarithmic amplifier of Fig. 6.19.

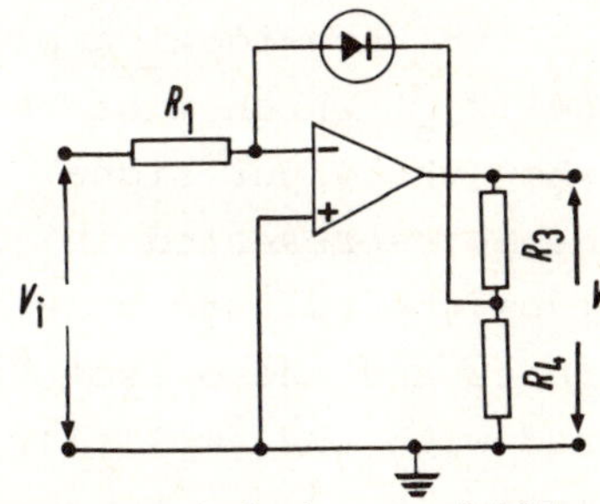

Fig. 6.21 A logarithmic amplifier with a scaling control.

provided the values of R_3 and R_4 are chosen so that $(R_3 + R_4)$ is not less than 2 kΩ.

The simple arrangement (Fig. 6.19) can be shown to perform well over four decades of current. Using the same principles, more complex circuits can be designed to give good performance over perhaps seven decades of current. An improved performance can be obtained by considering the relationship.

$$I = I_S \exp\,(eV/kT)$$

for a number of silicon diodes at constant temperature over the current range 10^{-2} to 10^{-8} A and selecting the most suitable device. Some manufacturers market diodes specially selected for use in logarithmic amplifiers.

6.2.16 *Measurement of the input bias currents and the offset voltage of an operational amplifier.* Although it is often assumed for simplicity in calculations that the operational amplifier is 'ideal', so the current flow into it is zero, in practice there are small input currents (called bias currents) of about 5 pA for a high quality amplifier, eg with dual fet inputs, and of about 30 nA (and so much easier to measure) for a low cost operational amplifier such as the 741.

In the equivalent circuit of the amplifier (Fig. 6.22) the input bias currents I_b are represented by the current generators.

Another feature of the non-ideal amplifier is the small input-offset voltage (V_{io}) which must be applied if zero is to be obtained when the input signal is zero. This input-offset voltage is represented in the equivalent circuit (Fig. 6.22) by the voltage generator V_{io}. The amplifier bias currents and the offset voltage are most conveniently measured with the amplifier in a closed loop configuration as in Fig. 6.22.

First, disconnect the 10 kΩ zero adjustment potentiometer (normally used to compensate for V_{io}, but not shown in the circuit of Fig. 6.22). The resistances R_1,R_2 and R_3 should be within 1 per cent of their stated values: a check on a standard resistance bridge is necessary if the their values are not known accurately. Measurements should be made with the amplifier supply voltages at +15 V and at -15 V. In the circuit of Fig. 6.22, when one of the switches S_1 or S_2 is open, bias current flows through the resistance R_3 and so an input voltage is applied.

With both switches S_1 and S_2 closed, the input voltage is V_{io} because the resistance R_1 is small enough for $R_1 I_b$ to be neglected.

Four readings (best made with a digital voltmeter) are necessary to measure the input bias currents and the offset voltage; they are as follows;

(i) With switches S_1 and S_2 closed, the output voltage V_{01} is measured where

$$V_{01} = (1 + R_2/R_1)\ V_{io}$$

(ii) With switch S_1 closed but S_2 open, the output voltage V_{02} is measured, where

$$V_{02} = (1 + R_2/R_1)\ (V_{io} + I_b^+ R_3);$$

(iii) With switch S_1 open but S_2 closed the output voltage V_{03} is measured, where

$$V_{03} = (1 + R_2/R_1)\ (V_{io} + I_b^- R_3)$$

(iv) With switches S_1 and S_2 open, the output voltage V_{04} is measured, where

$$V_{04} = (1 + R_2/R_1)\ |V_{io} + (I_b^- - I_b^+) R_3|$$

$$= (1 + R_2/R_1)\ (V_{io} + I_{io}\ R_3)$$

where the differential input-offset current

$$I_{io} = I_b^- - I_b^+ .$$

Typical results for two 741 amplifiers

(a) $V_{01} = +\ 0.76$ V, $V_{02} = -\ 2.8$ V, $V_{03} = +\ 5.1$ V,

$V_{04} = +\ 1.6$ V.

Using the above equations gives

$V_{io} = 7.6 \times 10^{-4}$, $I_b^+ = 35.6$ nA, $I_b^- = 43.4$ nA.

$I_b^- - I_b^+ = 7.8$ nA, $I_{io} = 8.4$ nA.

(b) $V_{01} = +\ 1.4$ V, $V_{02} = -\ 6.5$V, $V_{03} = +\ 11$ V,

$V_{04} = +\ 3.3$ V.

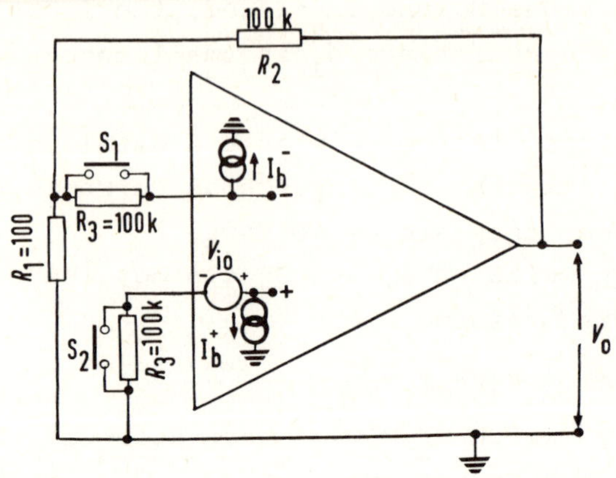

Fig. 6.22 Circuit for the measurement of input bias currents and offset voltage.

Calculations on the same lines are made from these values to give V_{io}, I_b^-, I_b^+, and I_{io}.

6.3 *Applications of operational amplifiers*

The operational amplifier is perhaps the most versatile device in modern electronics. There are so many possible applications of these devices in electronic circuits that it is inevitable that only a few can be considered below.

6.3.1 *Use of an operational amplifier to current-drive a meter.* The meter (eg a microammeter with fsd of 100 μA and resistance 1 kΩ) is used as a feedback path from the output to the input of the operational amplifier (Fig. 6.23). Assuming that no current enters the amplifier, the current through this meter in the feedback path is equal to the input voltage divided by the input resistance. With an input voltage of 10 mV and an input resistance of 100 Ω, the current recorded on the microammeter is 100 μA. This enables the resistance R_1 presented to the source to be considerably less than the resistance of a conventional microammeter. The meter is said to be 'current-driven' as from a high resistance source.

In the experiment with the circuit of Fig. 6.23,

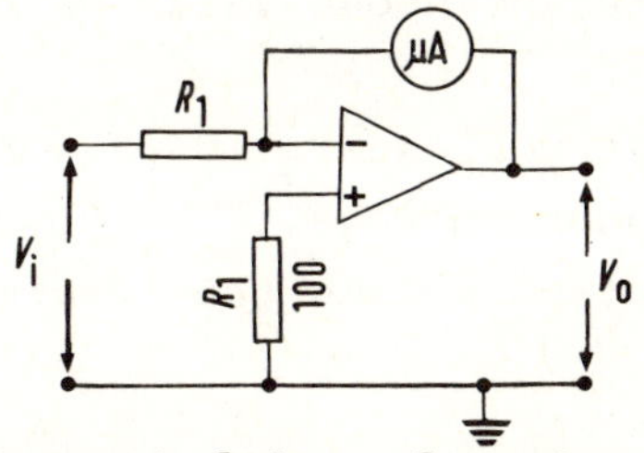

Fig. 6.23 Current drive of a microammeter by the use of an operational amplifier.

R_1 = 100 Ω, the input terminal is first connected to the common line and the voltage offset potentiometer is adjusted to give zero output voltage. A series of values of the steady input voltage V_1 in the range 1 to 10 mV (measured with a digital voltmeter or a potentiometer) is applied and the current recorded by the microammeter is noted at each value of V_1. The resistance R_1 should be known to within 1 per cent. A graph of this current reading against the input voltage is drawn and it is confirmed that the meter reading is equal to the current through R_1.

6.3.2 *The operational integrator and its use as a ramp generator*. The circuit of the basic operational integrator (Fig. 6.24) has a capacitor C as the feedback path from output to input of the operational amplifier. With an input signal v_i (which is a function of time) applied via the input resistance R_1, the instantaneous current i must charge the feedback capacitor C. As the output voltage v_o is given by

$$v_o = -(1/C)\int_0^t i\,dt$$

therefore

$$v_o = -(1/R_1C)\int_0^t v_i\,dt$$

where v_i is the instantaneous value of the input voltage at time t.

Thus, the output voltage v_o is proportional to the time integral of the input voltage. The time constant CR in this equation is often called the 'characteristic time' of the integrator.

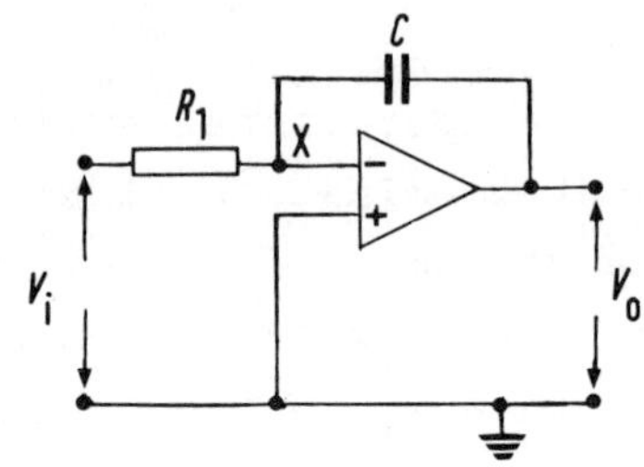

Fig. 6.24 The basic operational integrator

In the selection of an operational amplifier for use in an integrator, it is good practice to ensure a cheap unit is not used for which the input-offset voltage and the bias current may cause a continuous charging of the feedback capacitor so that, even when the input signal is zero, the output voltage drifts until the amplifier saturates.

In an experiment, a square wave generator is used to apply an input signal of 8 V peak-to-peak at a frequency of 5 kHz (Fig. 6.25). The input and output waveforms are displayed simultaneously on a double-beam cro. The output voltage (ie the voltage across the capacitor C) increases exponentially, but the observed portion is almost linear because the resistance in parallel with C is large, and only the initial portion of the charging cycle is being observed at a frequency of 5 kHz. To demonstrate that the output voltage is increasing exponentially with time, the 100 kΩ resistor is replaced by a smaller one of, say 10 kΩ.

This circuit (Fig. 6.25) is convenient to use because it is stable in operation and does not require a resetting facility. The larger the value of the resistance in parallel with the capacitor *C*, the more closely the circuit behaves as an ideal integrator.

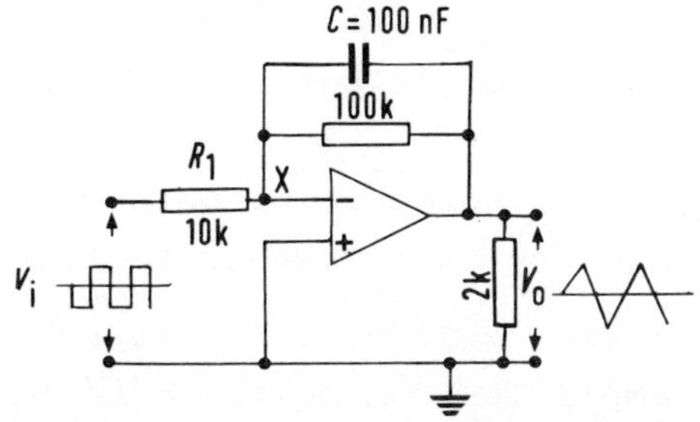

Fig. 6.25 Circuit utilizing an operational integrator to convert a voltage waveform of square shape to one of a linear triangular shape.

6.3.3 *Questions.*

(a) Is it possible to arrange for a linear ramp with a duration of 20 s by the use of high quality capacitors in an operational integrator circuit?

(b) How does the linearity of the ramp produced with the operational integrator compare with that obtained by means of a bootstrap circuit?

6.3.4 *A squaring circuit.* Set up the circuit of Fig. 6.26 based on an operational amplifier (eg type E78 or 741 to which the input is from a signal generator having a sinusoidal waveform output of peak voltage 1 V at a frequency of 1 kHz. Display the input and output waveforms simultaneously on a calibrated double-beam cro and record them. For the output signal measure and record the peak value, the pulse repetition rate (prr), the pulse duration and the pulse interval. The saturation condition

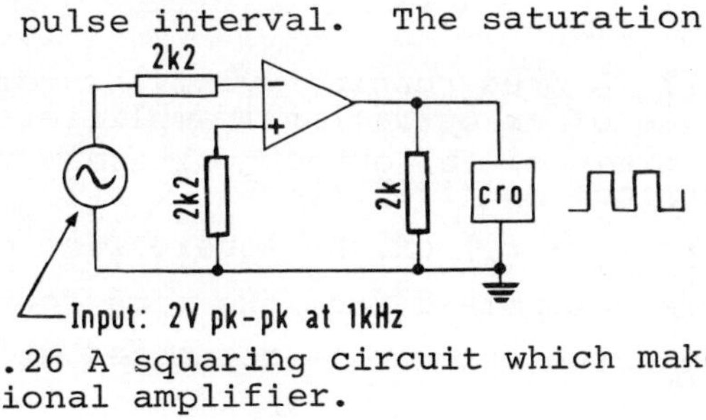

Fig. 6.26 A squaring circuit which makes use of an operational amplifier.

of the transfer characteristic of the operational amplifier is used in this 'squaring' circuit. The amplifier in Fig. 6.26 is operated on 'open-loop' and a positive and a negative-going input signal will switch the output from negative saturation to positive saturation at the signal frequency.

Investigate the performance of this circuit as the input signal frequency is increased to 10 kHz.

6.3.5 *A free-running multivibrator based on the use of an operational amplifier.* A suitable arrangement (Fig. 6.27a) is built around the operational amplifier 741.

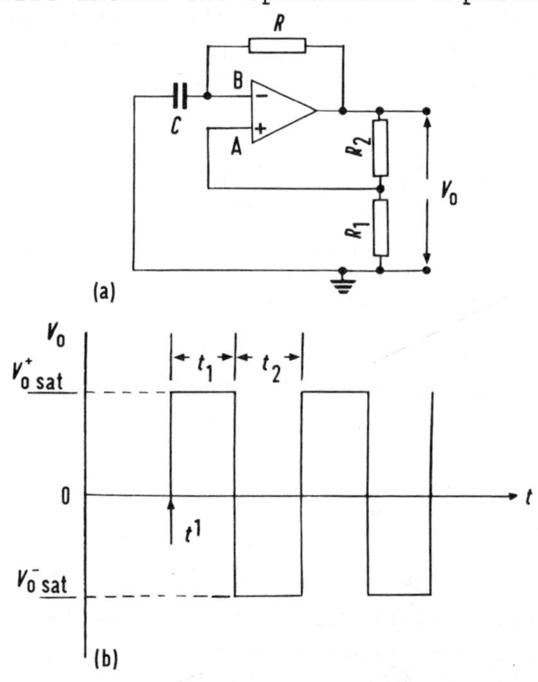

Fig. 6.27 A free-running multivibrator based on the use of an operational amplifier: (a) circuit diagram; (b) output voltage waveform from a symmetrical multivibrator.

The period T (= $t_1 + t_2$) of the square wave output voltage (Fig. 6.27b) is determined from the time constant CR. Assume that when the circuit is connected at time t^1

(Fig. 6.27b) the amplifier goes into positive saturation. The voltage at the non-inverting terminal A is then βV^{+}_{sat} where β, the feedback function, is $R_1/(R_1 + R_2)$. The inverting terminal B is negative in potential with respect to A and its potential rises as the capacitor C charges through the resistor R. When the potential at B is equal to, or slightly greater than that at A, the amplifier switches rapidly to negative saturation. During this rapid transition, the voltage at B remains virtually constant. Capacitor C now begins to discharge through R to approach βV^{-}_{osat}, when the next transition will occur. The time period T of the multivibrator, assuming a symmetrical output amplifier, is given by

$$T = RC \log_e (1 + R_1/R_2). \tag{6.14}$$

Set R_1 = 22 kΩ, R_2 = 47 kΩ, C = 100 nF and use a resistance box with its value set at 5 kΩ for R. Observe the output waveform with a cro and record it. Vary R between 1 kΩ and 10 kΩ and measure at each value the period T of the output voltage waveform from the multivibrator. T is best measured by means of a counter-timer or a digital frequency meter though a cro with a calibrated time-base can be used.

To verify the form of equation (6.14) plot a graph of the measured time period T_{obs} against the calculated time period T_{calc}. Repeat the experiment with $C \doteq 1\mu F$.

6.3.6 *Sensitivity of the frequency of the free-running multivibrator based on an operational amplifier to the supply voltage.* With a voltage supply to the operational amplifier of the circuit of Fig. 6.27 set at ±15 V and with the values of R_1,R_2,R and C given in Section 6.3.5. measure the period T of the multivibrator. Decrease the supply voltages V_S^{+} and V_S^{-} in steps of 0.5 V from ±15 V to ±12 V and at each step measure the period T (best done by measurement of the frequency f with a digital frequency meter, where $f = 1/T$). Over the voltage range used determine the magnitude of $(1/f)(df/dV)$ and express this as a percentage change per volt change of V_S.

6.3.7 *A monostable multivibrator based on an operational amplifier.* For constant values of R and C in the circuit of Fig. 6.28 the pulse width is determined from the magnitude of the reference voltage (V_{ref}) applied to the inverting terminal of the operational amplifier. With a negative reference voltage, the output from the monostable (in the stable state) is at positive saturation V^+_{osat}. The timing capacitor C is responsible for the regenerative switching which provides the fast transition. The negative voltage step at the output is transmitted via C to the noninverting terminal A which is therefore at voltage V^-_{osat} at the beginning of the semi-stable state. The voltage at A now rises exponentially as the capacitor C charges through the resistance R. The circuit will therefore revert to its stable state when the voltage at A reaches V_{ref}. The pulse width t is given by

$$t = CR \log_e |(V^+_{osat} - V^-_{osat})/V_{ref}| \tag{6.15}$$

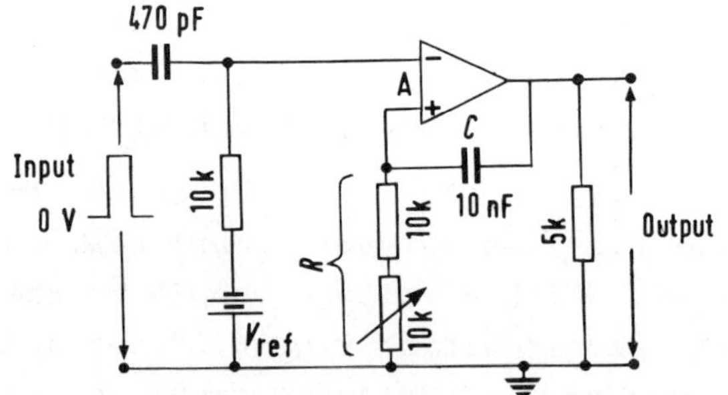

Fig. 6.28 A monostable multivibrator based on an operational amplifier

A suitable operational amplifier to use to build such a monostable multivibrator is either the 741 or the E78. The reference voltage V_{ref} is provided from either a stablilized voltage supply or dry batteries. With C = 10 nF, the transition is initiated with a positive-going signal from a pulse generator and the input and output waveforms are displayed on a double-beam cro. The output pulse width is varied by altering the value of R. Measurement of the output pulse width using known values of C,R and V_{ref} enables a measured value of t to be compared with that calculated from equation (6.15).

6.3.8 *Voltage stabilization based on an operational amplifier.* An operational amplifier can be used in voltage follower mode (with unity gain) to provide a means of high stabilization of a voltage supply. In an experiment (Fig. 6.29) the differential input operational amplifier E78 is used: This has a rated dc open-loop gain of 10^5; a standard cell (SC) in the form of a standard mercury cell or a dry cell is used to provide the input voltage between the non-inverting terminal and earth. In voltage mode the current provided by the amplifier will arrange to keep the potential difference between the inputs equal to zero, so that the voltage across the standard cell terminals will appear across the load resistor comprising R_1 and R_2 in series. The fixed value resistor R_1 is usually chosen to be 1 kΩ so that the current provided by the amplifier does

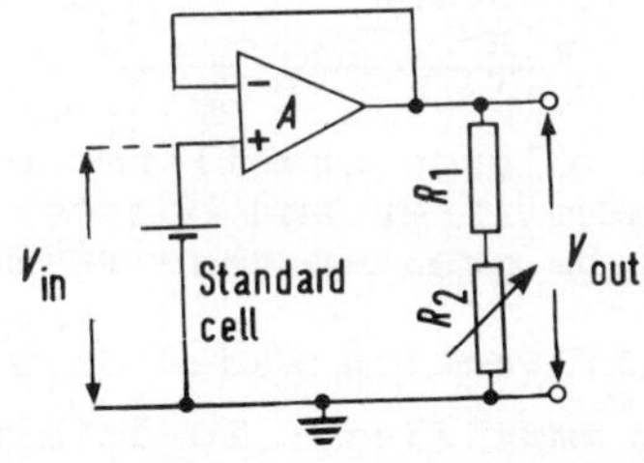

Fig. 6.29 A voltage stabilizer circuit based on the use of an operational amplifier in voltage follower mode.

not exceed its rated value of 10 mA. If voltages exceeding that of the standard cell are required, the circuit of

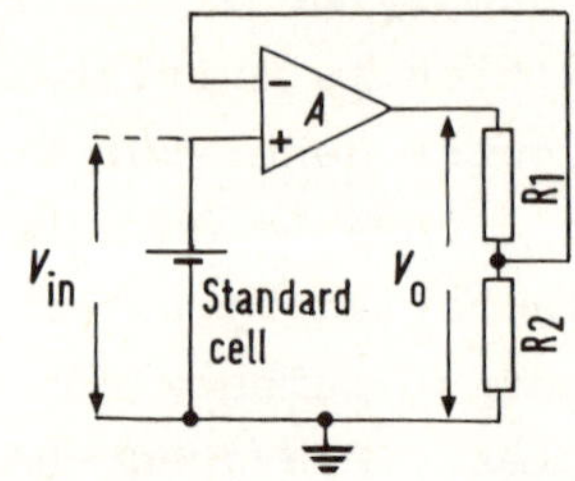

Fig. 6.30 A voltage stabilizer circuit with an operational amplifier connected so that the output voltage can exceed the standard cell voltage V_{in}.

Fig. 6.30 can be used, whereas if currents greater than 10 mA are needed, a booster amplifier (Fig. 6.31) is used.

To determine the stabilization factor S_V the supply voltage to the amplifier is varied between, say +12 - 0 - -12V and +15 - 0 - -15V and any resting output voltage change is recorded. S_V and R_o (see Sections 3.7 and 3.7.1) are determined for load currents of 1 mA and 5 mA. The current provided by the operational amplifier E78 must not be allowed to exceed 10 mA.

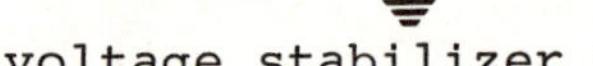

Fig. 6.31 A voltage stabilizer circuit with an operational amplifier and booster amplifier to enable currents greater than 10 mA to be supplied.

6.3.9 *Voltage stabilization based on a Darlington pair and an operational amplifier.* To determine the stabilization factor S_V, the stabilizer unit (Fig. 6.32) has its input terminals connected across a dc power supply source (as shown, a bridge rectifier with smoothing capacitor). Across the output terminals 00 is connected a 250 Ω (max) variable resistance set at half its maximum value in series with a milliammeter of fsd 150 mA. These two components (125 Ω plus the milliameter resistance) constitute the load resistance R_L. The input voltage V_i is adjusted to 20 V and the output voltage control set to provide an output voltage V_o of 12 V. The rheostat (250 Ω max) is adjusted until the load current I_L (as recorded by a series 0 - 250 mA meter) is 100 mA. Keeping I_L constant at 100 mA, the input voltage V_i is varied in steps of 1 V over the range 16 to 24 V and the output voltage V_o is recorded at each step. The change in output voltage is exceedingly small, as would be expected of a high performance voltage stabilizer. If the voltage V_o were found to change by less than 10 mV for a change in V_i of 8 V (16 V → 24 V),

$$S_V < 10^{-2}/8$$

ie $$S_V < 1.25\times10^{-3}$$

The experiment should be repeated with I_L maintained at 500 mA (this requires changing both the rheostat and the milliammeter which together constitutes the output load R_L).

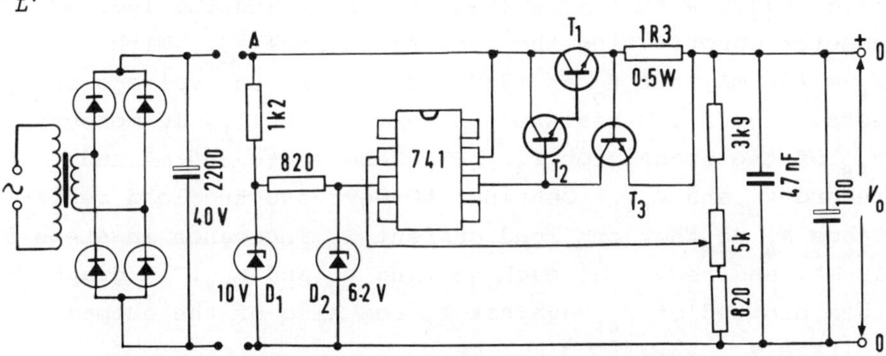

Fig. 6.32 A voltage stabilizer having a Darlington pair (T_1 and T_2 comprising a Darlington pair) as a series control element and an operational amplifier (741) as a difference amplifier.

To determine the output resistance R_o, the circuit

and its adjustments are as before and the rheostat (250 Ω max) is adjusted until the load current I_L = 250 mA a small adjustment to the output voltage control then being needed to ensure that the output voltage is still 12 V. By means of this rheostat the load current I_L is set at 50 mA intervals between 50 and 250 mA. At each of these current settings the output voltage V_o is measured. The 250 Ω (max) rheostat and 250 mA milliameter are then replaced by a 50 Ω (max) rheostat in series with an ammeter reading 0 to 1A. Now I_L is increased in steps of 50 mA between 300 and 500 mA, recording the output voltage V_o at each setting.

Once again, it is to be expected that the change in V_o would be small. In a typical experiment it was found that V_o decreased by 10 mV as the load current increased from 50 mA to 500 mA, giving

$$R_o = \left|\frac{\Delta V_o}{\Delta I_L}\right|_{V_i,T} = \frac{10^{-2}\Omega}{450\times10^{-3}} \simeq 22\ \text{m}\Omega$$

Limiting the output current. Set up this same circuit (Fig. 6.32) with a 50 Ω (max) rheostat and the (0-1 A) ammeter constituting the load resistance R_L. With I_L = 250 mA, set V_o at 12 V and measure the voltage drop across the 1.5 Ω resistor in series with T_1, ie measure V_{BE} of the transistor T_3. Increase I_L to 300 mA and record V_o and V_{BE}. Continue to decrease the load resistance R_L so that the load current I_L increases in steps of 50 mA, and record at each setting V_o and V_{BE}. A graph is then plotted of V_{BE} against I_L and also of the output voltage V_o against I_L.

6.3.10 *Questions.*

(i) What is the advantage of using a Darlington pair (T_1 and T_2 of Fig. 6.32) as the series control element?

(ii) How is the output current limited in this stabilizer circuit?

(iii) Calculate the power dissipated in the series control element when V_i = 20 V, V_o = 12 V and I_L = 500 mA.

(iv) Calculate the power dissipated in the 6.2 V Ze Zener diode of Fig. 3.8.

(v) This stabilizer circuit is called 'hybrid'. Why should this term be used?

6.3.11 *Further investigation.* With the unit providing 500 mA at 12 V, observe by means of a cro any fluctuations of the voltage across R_L. Any voltage fluctuation (other than 100 Hz ripple) is 'noise'. Measure the amplitude of any noise at the output.

6.4 *Voltage-to-frequency converters which make use of an operational amplifier*

There are a number of voltage-controlled oscillator (VCO) circuits based on various active devices in which an operational amplifier can be used with advantage. A few of these are described in the experiments which follow.

6.4.1 *A voltage-to-frequency converter based on a unijunction transistor.* In the circuit (Fig. 6.33) the capacitor C in the ujt-based relaxation oscillator (see Fig. 5.3) is charged from a constant current source (based on the bipolar transistor BCY70) so that a linear relationship exists between the voltages across C and the oscillator frequency. The operational amplifier shown in Fig. 6.33 (a 741 or E78 is suitable) is connected as a voltage follower with the supply voltages to this amplifier V_{CC} = $\pm$15 V. With a negative input voltage V_i applied to the non-inverting terminal, a constant current flows through the resistor R_3 to establish a pd across R_3 equal in magnitude to V_i. If the charging current to C is constant the rate of rise of the voltage on the emitter of the unijunction transistor is constant and the time taken for this current to equal the peak-point voltage V_P

is proportional to V_i. Hence, the frequency at which the ujt is switched to a conducting state is proportional to V_i so that a linear relationship is established between voltage and frequency.

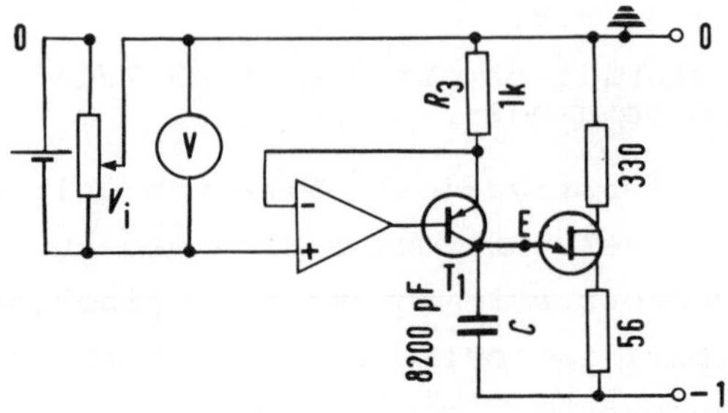

Fig. 6.33 A voltage-to-frequency converter based on a unijunction transistor circuit with an operational amplifier.

In an experiment based on the circuit of Fig. 6.33, the ujt is conveniently a 2N2646. The input voltage V_i is set at -1 V and the voltage waveform at the emitter E is displayed and recorded on a cro. V_i is varied and measured (preferably with a DVM) within the range 0.3 to 1.5 V and the frequency is measured (preferably with a digital frequency meter) at each setting. A graph is plotted of voltage against frequency.

Note that, as V_P for the ujt is given by

$$V_P = \eta V_{BB} + V_D$$

(see Section 5.1), the frequency of oscillation of the circuit of Fig. 6.33 can be raised by alteration of the interbase voltage V_{BB}.

To investigate this experimentally, set V_i at -2 V with the supply voltage V_{CC} to operational amplifier at ±15 V and adjust the voltage across the ujt circuit until the output frequency is 20 kHz. With the ujt voltage constant vary V_i within the range -0.5 to -2 V and plot a graph of voltage against frequency. Use the best straight line through the experimental points to calculate the constants in the linear equation

$$V_i = mf + C.$$

6.4.2 *A voltage-to-frequency converter based on a programmable unijunction transistor (put).* The use of a put (see Section 5.4) enables an improved version to be made of the circuit of Fig. 6.33. In an experiment, use the circuit of Fig. 6.34 with $R_1 = R_2 = 4.7$ kΩ so that $\eta = R_2/(R_1 + R_2) = 0.5$. With the input voltage $V_i = -100$ mV, display and record with a cro the voltage waveform across the capacitance C. Set values of V_i in the range -20 mV to -200 mV and measure them, preferably with a DVM; at each of these settings measure the pulse repetition rate with a digital frequency meter and plot a graph of voltage against frequency. Alter the values of R_1 and R_2 to 1.5 kΩ and 4.7 kΩ respectively so that $\eta = 0.75$ approximately and plot a second graph of input voltage against frequency.

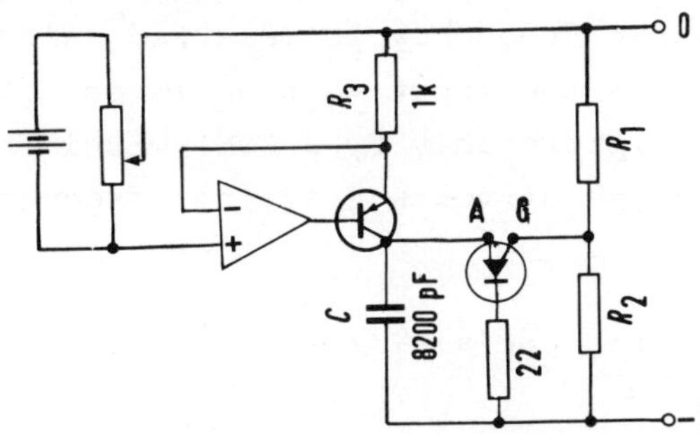

Fig. 6.34 A voltage-to-frequency converter based on a programmable unijunction transistor (put) with an operational amplifier.

6.4.3 *A voltage-to-frequency converter based on a free-running multivibrator utilizing field-effect transistors.* A circuit of this type is described in Section 4.4.3 in which bipolar transistors are used to act as constant current sources. To avoid the sensitivity to temperature change of bipolar transistors used in this way, operational amplifiers (eg type 741) serve excellently.

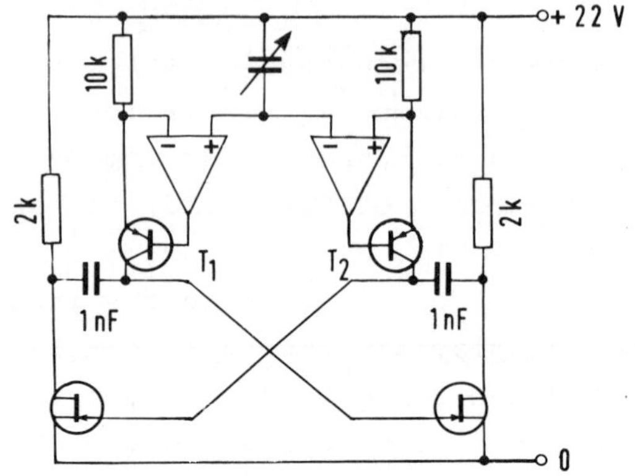

Fig. 6.35 A voltage-to-frequency converter based on a fet free-running multivibrator which makes use of two operational amplifiers as constant current sources; (bipolar transistors T_1 and T_2 are, eg each BCY 70).

A suitable circuit to set up an experiment is as shown in Fig. 6.35, and makes use of two 741 operational amplifiers. The output frequency is measured using a digital frequency meter as the input voltage is set at various measured values (preferably by a DVM) within the range 0 to 10 V. A graph of voltage against frequency is plotted.

6.5 *A high quality pre-amplifier for audio-frequency signals*

The demand is considerable for an amplifier system which operates with an input signal at audio-frequency. This input signal is usually of small amplitude from a transducer. The quality of the amplifier system is expressed in terms of its ability to provide at its output an amplified version of identical waveform to that of the input signal and where this output is at a power level sufficient to drive a loudspeaker, a control motor or a large visual display. The basic requirements are much the same whether it is to be used in a measuring, a control or a sound reproduction system. The amplifier system adopted in almost all such cases composes a pre-amplifier which enlarges the small current or voltage input and drives a following power amplifier the output of which operates the electromechanical device (eg a loudspeaker) in question. The object here is to study experimentally a pre-amplifier of high quality based on bipolar transistors. Perfection in an amplifier does not exist. As in all scientific work there is selected, in terms of performance and cost, a system which is adequate for the task.

Some of the terms used to specify the performance of an amplifier.

(i) Frequency response: the variation with frequency of the gain of the system. The 'ideal' amplifier would have a constant gain from zero frequency to an infinitely high frequency. Such a response is said to be 'flat', which is indicative of the shape of the graphical plot of the gain against the frequency.

(ii) Bandwidth: the width of the band of frequencies over which the power amplification does not fall below a specified fraction, usually one half or 3dB of the chosen value. In the case of voltage amplifiers, the bandwidth is often quoted as the frequency range over which the voltage gain does not fall below one-half of its mid-band value.

(iii) Noise: a term conveniently used to denote any features existing in the output signal which are not present in the input signal.

(iv) Distortion: any change in the waveform which results from the passage of the signal through the amplifier system.

(v) Efficiency: because the input signal is usually one of negligible power, the useful power at the output of an amplifier must be provided from the dc power supply.

The efficiency is the ratio of the useful output power to the power supplied to the amplifier. Invariably efficiency is sacrificed for quality. Such sacrifice is generally only of concern with portable equipment operated from batteries or solar cells.

(vi) High-fidelity (Hi-Fi): essentially a trade term used indiscriminately by advertisers to describe sound-producing equipment with quality ranging in practice from inferior to excellent. Attempts are now being made by manufacturers to specify minimum performance figures for equipment so described. The pre-amplifier to be studied (Fig. 6.36) makes use of discrete components and is arranged so that these components can be interchanged readily. Although pre-amplifiers of similar quality are available in integrated circuit form, they are not so convenient for study.

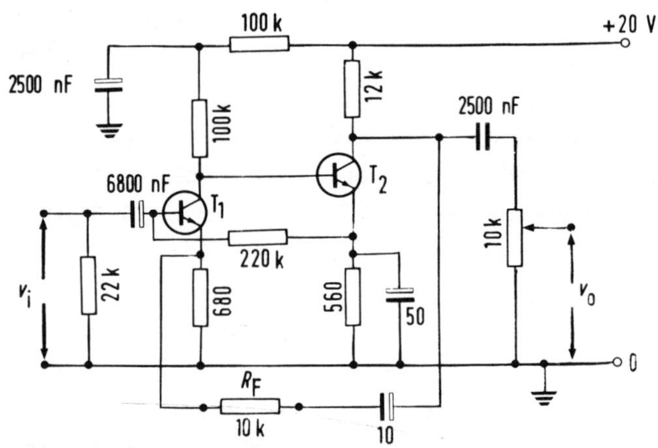

Fig. 6.36 A quality pre-amplifier for audio-frequency signals. (T_1 is, eg a BC109; T_2 is, eg, a BC108).

Features of the circuit of Fig. 6.36

(a) T_1 is a low-noise bipolar transistor

(b) The two stages are directly coupled with the advantage that there is no frequency-dependent passive component (such as a capacitor) used to interconnect them.

(c) Negative feedback is incorporated by, firstly, not having a by-pass capacitor across the 680 ohm resistor in the emitter lead of transistor T_1 and, secondly, by providing to the emitter of T_1 a feedback path for a fraction of the alternating output signal, this path comprising the 10 kΩ resistor R_F in series with a 10 μF capacitor. The effect on the amplifier of this ac negative feedback is to reduce the voltage gain, increase the input impedance and the bandwidth, reduce distortion and to render the performance of the amplifier less dependent on the characteristics of transistor T_1.

(d) The steady bias for transistor T_1 is taken from the emitter of T_2. This dc negative feedback stabilizes the operating point.

6.5.1 *The frequency response of the audio-frequency amplifier.* By means of a suitable signal generator (of which the output can be varied in amplitude and in frequency) a signal of 5.0 mVrms (measured by an ac millivoltmeter) at a frequency of 1 kHz is applied to the input terminal of apparatus constructed in accordance with the circuit diagram of Fig. 5.36. The input and output signals are recorded on a double-beam cro. One would expect the two signals to be in-phase as each stage of the amplifier produces a phase-change of 180°. With the output signal from this amplifier set at its maximum value (the 10 kΩ potentiometer could serve as volume control in an audio-system), measure the input and output signals by means of the same ac millivoltmeter. Repeat these readings at, say, 20 different frequencies over the range from 10 Hz to 1 MHz. Tabulate these readings. Calculate the voltage gain $A_V = v_o/v_1$ and express this voltage ratio in decibels (dB). Plot a graph of the gain in dB against the frequency, a typical result being that shown in Fig. 6.37. Calculate the bandwidth of this amplifier on the basis that it is the frequency band over which the voltage gain does not drop below 1dB of its mid-band value. Note that this pre-amplifier can extend considerably the use of a low-cost cro.

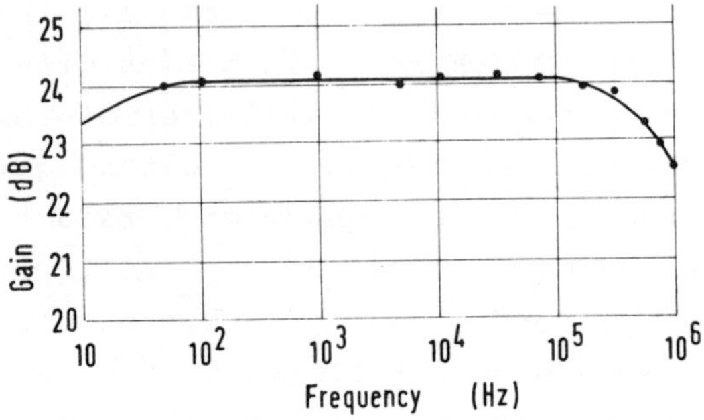

Fig. 6.37 Typical graph of the gain in dB against the frequency for the quality pre-amplifier.

6.5.2 *The dependence of the amplifier voltage gain on the supply voltage.* To study this experimentally and to show the extent to which this gain is not dependent on the supply voltage, set up the circuit of Fig. 6.36, with an input signal at 5 mVrms at a frequency of 5 kHz. With the output signal set to its maximum value, record the output voltage by use of the ac millivoltmeter. Vary the supply voltage to the amplifier in steps of 0.5 V over the range from +18 V to +22 V and measure the voltage gain at each setting.

6.5.3 *The dependence of the amplifier performance on the characteristics of the individual transistor.* The previous experiment (Section 6.5.2) gives the voltage gain of the amplifier (circuit of Fig. 6.36) for a 5 mV input signal at 5 kHz. Interchange the transistors T_1 and T_2 and again measure the voltage gain. With the transistor T_2 as a BC108, replace the original BC109 for transistor T_1 with one of lower current gain and record the voltage gain of the amplifier.

6.5.4 *An equalization network.* In the recording of sound one method involves cutting into the material of the disc a lateral groove of amplitude inversely proportional to the frequency of the sound. Thus bass notes of low frequency cause relatively large lateral swings of the cutting needle tip whereas treble notes of high frequency cause very small lateral swings. To avoid excessive lateral movement of the cutting stylus which could result in the stylus cutting into an adjacent groove, the low frequency signals are attenuated relative to the high frequency ones when recording. On playback, the original relative amplitudes are restored by amplifying the bass notes more than the treble ones. This tone modification is known as 'equalization'. Several equalization characteristics are available based on the recommendations of professional societies, for example, the Recording Industry Association of America (RIAA). The equalization network shown in Fig. 6.38, used in place of R_F in the circuit of Fig. 6.36, is designed to produce a frequency response in accordance with RIAA regulations for use with magnetic cartridges.

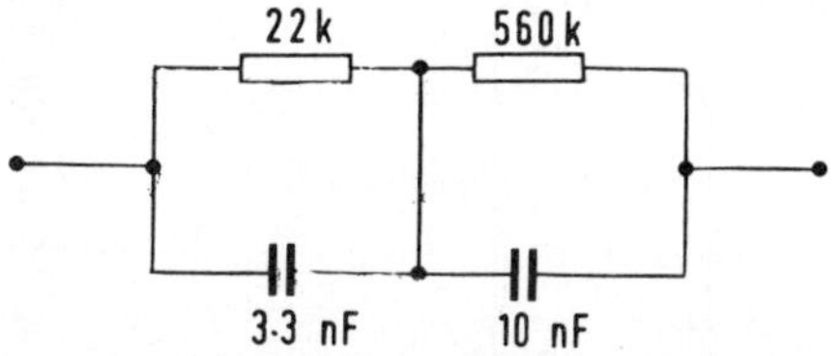

Fig. 6.38 Equalization network to use in place of R_F in the amplifier circuit (Fig. 6.36) to produce a frequency response in accordance with RIAA regulations for use with magnetic cartridges.

The experiment is to replace R_F by the network shown in Fig. 6.38 and measure the gain of the amplifier in the way described in Section 6.5.1 The gain at a frequency of 1 kHz is used as the reference and is termed OdB. The relative response, expressed in dB, is calculated with

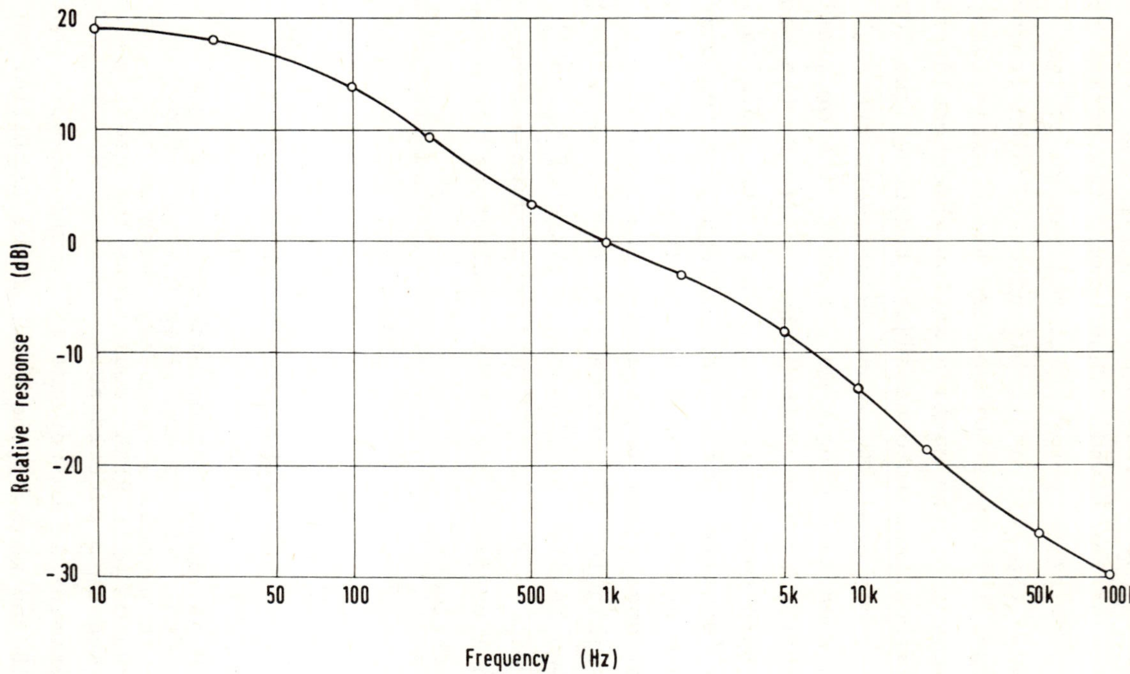

Fig. 6.39 Typical graph of relative response against frequency on using the equalization network (Fig. 6.38) in place of R_F (Fig. 6.36)

respect to this reference. Plot a graph of relative response against frequency; a typical result is shown in Fig. 6.39. This plot shows that, relative to the mid-band reference frequency of 1 kHz, the bass notes are amplified whereas the treble ones are attenuated. When the pre-amplifier is used with a tape recorder, different equalization networks are switched in.

6.5.5 *Further investigations.* (a) Construct the circuit shown in Fig. 6.36 on Veroboard; (b) plan the layout of and fabricate a printed circuit board for the pre-amplifier.

6.5.6 *The input impedance of the pre-amplifier.* Connect a resistor of 10 kΩ (1% type) into the base lead of the transistor T_1 as shown in the circuit of Fig. 6.40. Apply a sinusoidal input frequency of 1 kHz from a signal generator so that the voltage v_1, as recorded by an ac millivoltmeter, is 5 mVrms. Measure the voltage v_2 and calculate the base current $i_p = (v_1 - v_2)/10^4$. The input impedance of the amplifier is therefore given by v_2/i_b and should be of the order 150 kΩ. It must be confirmed that the ac millivoltmeter used has a high impedance so that, when recording v_2, it does not add appreciably to the potential drop across the 10 kΩ resistor.

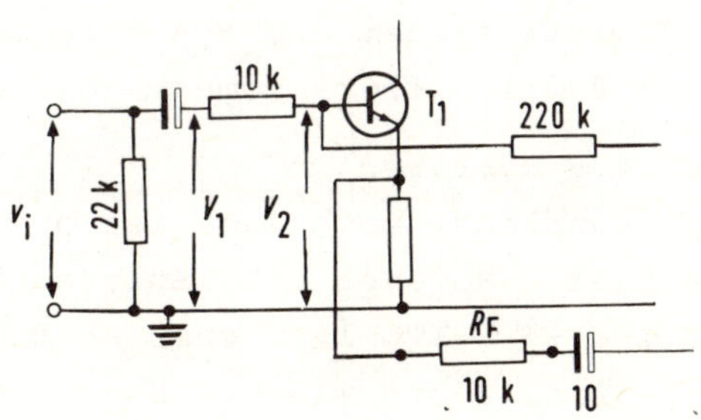

Fig. 6.40 Determination of the input impedance of the quality pre-amplifier.

CHAPTER SEVEN

Logic gates

7.1 *Introduction*

Devices used in logic circuits to control or to effect the passage of information through the system are called logic gates. There are two widely-used families of logic gates, both in integrated circuit form: (a) transistor - transistor logic (TTL) and (b) complementary - metal - oxide - semiconductor (CMOS, pronounced 'see-moss'). Both are to be described here but with attention given first to TTL because it came first chronologically and has been used more, although CMOS is gradually replacing TTL in many applications. Typical examples of these two families are the 7400 TTL series (introduced by Texas Instruments) and the 4400 CMOS series. While the fundamental principles involved in a digital system are the same and must be understood, each family offers some major advantages.

7.1.1 *TTL and CMOS compared.*

(a) Voltage supplies: both families operate from a single voltage rail. However, the tolerance is only about 5 per cent for the +5 V TTL line whereas for CMOS the tolerance can be from +3 to +18 V. Hence CMOS (unlike TTL) will work reliably from an unregulated and poorly filtered voltage supply. Special CMOS devices which have been developed for digital watches operate on a +1.0 V line.

(b) Power consumption: for a CMOS system, power

consumption depends on operating speed: operating continuously at 2 MHz a typical CMOS logic gate, working from a 5 V line, will consume 1 mW approximately whilst reduction of the frequency to 20 kHz reduces the power demand to as low as 0.01 mW. On the other hand, for a TTL gate, the power consumption will be 1 mW at both frequencies so that frequency reduction favours considerably the CMOS gate with regard to power consumption. Moreover in a quiescent state, the power consumption of a CMOS gate is of the order of nW whereas for a TTL gate, it is of the order of mW: again CMOS is better.

(c) Noise immunity: spurious voltage spikes (noise) occur frequently on some types of line connection and may well cause premature switching of logic gates. The noise immunity of CMOS gates exceeds greatly that of TTL ones. For the CMOS gate a noise immunity of 4 per cent of the supply voltage is typical.

(d) Packing density: CMOS is again favoured because a bipolar active device in TTL occupies about 20 times the space needed for an individual mosfet in CMOS. Consequently CMOS lends itself much better than TTL to large scale integration (LSI).

(e) Operating frequency: TTL gates are superior to CMOS because they are much less capacitance limited and can hence operate at frequencies as high as 100 MHz. For integrated circuit CMOS gates, the maximum operating frequency is about 10 MHz because, during each switching transient, the gate and interconnection capacitances have to be charged or discharged via a path of large resistance.

7.1.2 *Basic constraints in TTL logic circuit design and operation.* Among these are the following.

(i) The signal level at both input and output must occupy one of two states. In one state, designated by '0' or 'logical 0', the case of TTL is typically +0.2 V, (+0.4 V max). In the other state, designated '1' or 'logical 1', the voltage level is typically +3 V (+ 2.4 V min). Positive logic is involved when the voltage used to represent logical 1 is more positive than that used to represent logical 0 whereas negative logic prevails if these levels are interchanged.

(ii) In the case of TTL, the signal level should never exceed +5.5 V or be less than 0.6 V.

(iii) In the case of TTL, the time taken for the transition from logical 0 to logical 1 or vice versa (the rise and fall times respectively) should not exceed 1 μs. This is because, as the circuit switches from one stable state to the other, it passes through an unstable region where it may oscillate.

Any circuit which operates with these constraints will maintain the appropriate voltage levels when connected either to the input or to the output of a TTL gate and is said to be 'TTL compatible'.

7.1.3 *A logic probe.* A convenient method of testing the state (voltage level) of any pin in a system of logic gates is essential. This is provided, for example, by a logic probe (Fig. 7.1) designed to operate with 5 V TTL logic gates. This probe is housed within a plastic tube, is powered from the available +5 V logic-circuit supply voltage, and contains two light-emitting diodes (LEDs) A and B. When the fine conducting probe is in contact with the pin (of the integrated logic circuit package) under investigation, LED *A* will emit light only if the voltage level is in the 'low' or '0' state, while LED *B* will emit light only if the voltage level is in the 'high' or '1' state. If neither A nor B emits light, the voltage level may be outside the limits for TTL gates or the pin is open-circuit. The ability of the probe to identify these three distinct states (Fig. 7.2) is particularly useful. Because pulses of very short duration will not be visible it is recommended that during a test a minimum pulse duration of 100 ms is used.

More complex logic probes incorporate a circuit which extends the pulse width so enabling an input pulse of very short duration to be detected. Such a circuit is often called a 'pulse stretcher'.

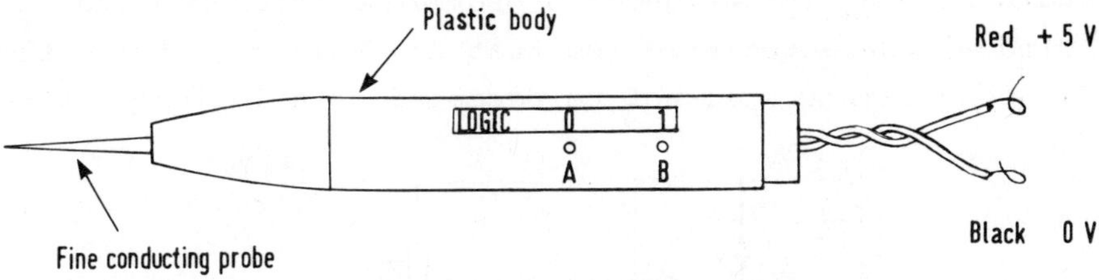

Fig. 7.1 A logic probe for TTL gates

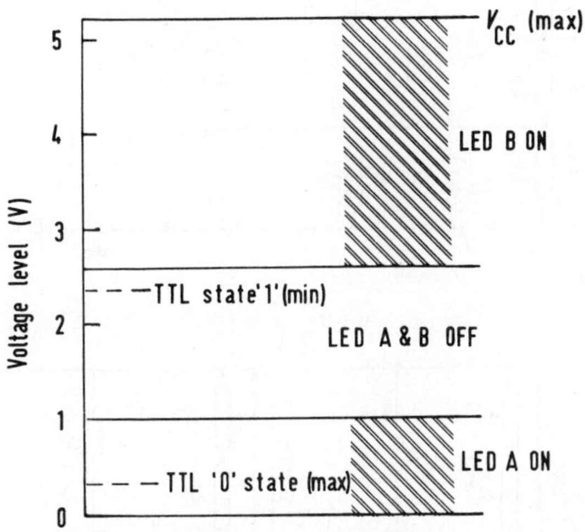

Fig. 7.2 Logic probe characteristics suitable for TTL gates.

Fig. 7.3a is the circuit diagram of the logic probe illustrated by Fig. 7.1. Each state acts as a discriminator and driver for the indicating LED. With the probe leads connected between +5 V and 0 V of the supply to the logic gate and the probe open-circuit, transistors T_1 and T_2 are both biased in an OFF state. With the probe sensing a logical 0 state (Fig. 7.3b) the current I_{sink}

passes through the resistors R_1 and R_2 to render transistor T_1 conducting so that LED *A* will glow. With the probe sensing a logical 1 state, the current I_{source} flows through the resistors R_3 and R_4 so that transistor T_2 is rendered conducting and LED *B* will glow. The resistor chain $R_1 \rightarrow R_4$ is designed to provide a threshold and 'window' characteristic so that any voltage outside the TTL constraints will not be accepted as a logical 0 or

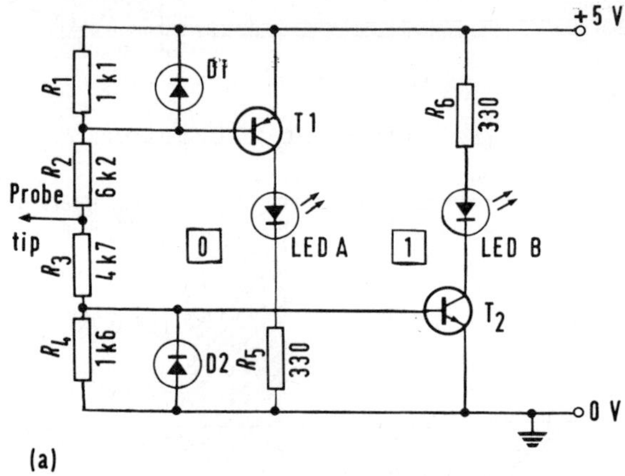

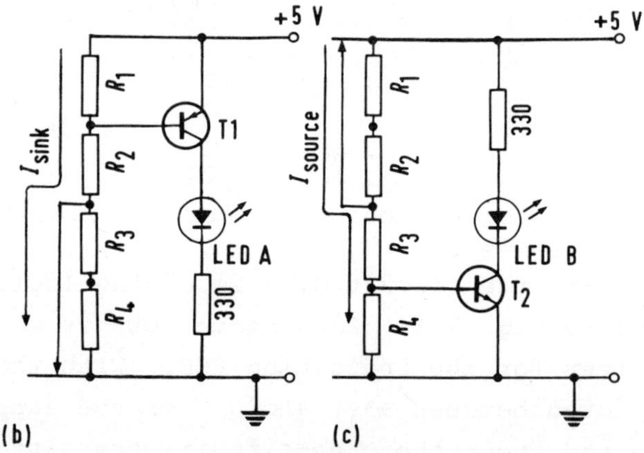

Fig. 7.3 A logic probe: (a) circuit diagram; the voltage supplies must be those serving the gate under investigation; (b) detecting a logical 0 state; (c) detecting a logical 1 state.

logical 1 state. The diodes D_1 and D_2 protect transistors T_1 and T_2 respectively against any excessive reverse voltage if the probe tip inadvertently touches a point at a potential above 5 V or below 0 V.

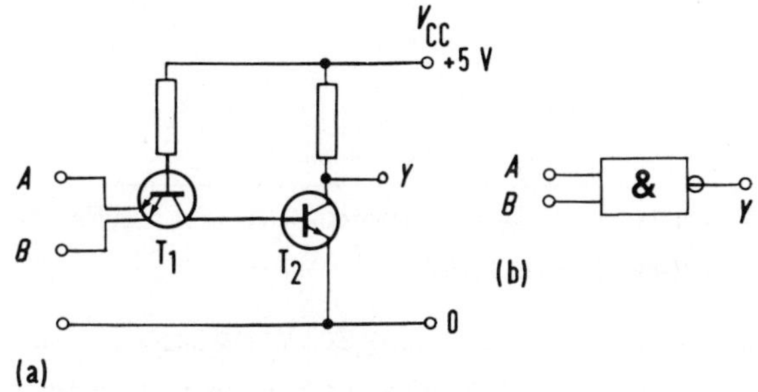

Fig. 7.4 A basic 2-input TTL NAND gate: (a) circuit diagram; (b) the logic symbol for a NAND gate.

7.2 *The basic TTL 2-input NAND gate*

The circuit is shown in Fig. 7.4a whereas the accompanying symbol (Fig. 7.4b) is that of the NAND gate. If all the inputs to the multi-emitter transistor T_1 (3 and 4 inputs are possible) are at the logical 1 level, all the base-emitter junctions of T_1 are reverse-biased. However, the base-collector junction is forward-biased and provides the base current which drives transistor T_2 into saturation. The output voltage in this state is therefore at logical 0. If any one input goes to logical 0, the corresponding base-emitter junction becomes forward-biased and conducts heavily. The voltage between the base and the emitter of transistor T_2 is then insufficient to make it conducting. With T_2 cut off, the output voltage rises to logical 1.

The performance of the NAND gate is illustrated by the truth table (Table 7.1). The AND function is included as the intermediate step.

Table 7.1 Truth table for a 2-input NAND function

Input		Output	
		AND	NAND = $\overline{\text{AND}}$*
A	*B*	*Y*	*Y*
0	0	0	1
0	1	0	1
1	0	0	1
1	1	1	0

* NAND = NOT AND; the bar (above AND in this case) denotes the NOT or INVERT function.

7.2.1 *The SN 7400N quad 2-input NAND unit.* A 14-pin dual in-line package suitable for experiments contains 4 separate 2-input NAND gates: it is one member of the extensive 7400 TTL system of logic circuits identified as type SN 7400N, and is called 'a quad 2-input NAND unit'. When this unit is plugged into its socket, the recess on the top surface of its plastic body is always aligned with a similar recess in the socket. An enlarged diagram on the module shows the device viewed from above; with the recess on the left-hand side, pin number 1 is in the bottom left-hand corner. Each of the four NAND elements within the SN 7400N has two input terminals labelled *A* and *B* and one output terminal labelled *Y*.

In experiments with this SN 7400N to study the state of the output, it is convenient to make use of an n-p-n transistor (eg type BC 109) which has in its collector lead a light-emitting diode (LED) in series with a 220 Ω resistor and a 1 kΩ resistor in its base lead (Fig. 7.5). With +5 V applied at point P (ie to the transistor base via the resistor) the output is high (at the level of logical 1) and light is emitted from the LED; when the voltage applied at point P is less than +0.4 V the output is low (at logical 0) and the LED does not emit light.

To investigate the NAND function set up the circuit of

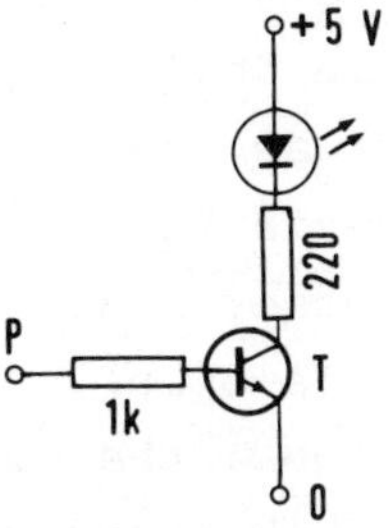

Fig. 7.5 Circuit used to investigate the output state of a TTL unit.

Fig. 7.6. Note that the voltage supplies are not shown in logic diagrams: it is assumed that the necessary supplies are connected; in the case illustrated in Fig. 7.6 pin 14 of the SN 7400N is connected, in fact, to +5 V and pin 7 is earthed. Confirm that when any input to a NAND gate is floating, the gate assumes a logical 1 level. Hence, with both inputs floating the output should be at logical 0. Now connect *A1* to earth and observe the output level with the LED. Continue in this way to verify the information given in the truth table (Table 7.1).

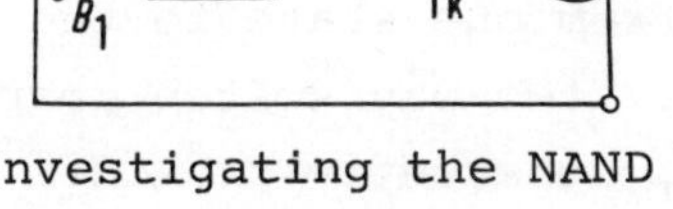

Fig. 7.6 Investigating the NAND function

Connect together the two inputs of the NAND gate; now it behaves simply as an inverter, a logical 1 at the input appearing as logical 0 at the output and vice versa.

7.3 *Multivibrator circuits based on NAND gates of the TTL type*

Multivibrator circuits as considered in Section 3.10. Three basic classes of this type of circuit are termed free-running, monostable and bistable. All three depend on the use of two active devices (bipolar transistors in Section 3.10) which act as 'switches' in that in one state the active device is full conducting (ie the switch is closed) and in the other state the active device is cut-off (ie the switch is open). In Section 3.10, and the following sub-sections, the circuit diagrams of Figs. 3.20, 3.23, and 3.26, each based on bipolar transistors T_1 and T_2, can be considered alternatively as coupled switches S_1 and S_2 where for the free-running multivibrator, S_1 is coupled to S_2 by means of an *RC* circuit and S_2 is coupled back to S_1 by a similar *RC* circuit; for the monostable circuit S_1 is coupled by *RC* to S_2 but the reverse coupling from S_2 to S_1 is direct (without the use of a capacitor; only a potential divider resistor arrangement is used) and for the bistable circuit direct coupling is present both ways: from S_1 to S_2 and vice versa.

It is clear that on NAND gate with two inputs can be used as a fast-acting switch and hence that multivibrator circuits can be based on them.

7.3.1 *A bistable circuit based on the use of two 2-input NAND gates.* It is often required to produce a steady voltage corresponding to the logical state 1 or 0 and to be able to switch from one state to the other, as required. A simple mechanical, two-way switch operating between, say, +5 V and 0 V , apparently provides the answer and so could be used to provide the needed logic-level input voltages to logic gates. There is, however, an important disadvantage in this simple solution to the problem: such a mechanical switch rarely produces a clean transition from one voltage level to the other because, when the mechanical contact is broken, several pulses occur before the steady state is reached: these fast spurious pulses are said to be the result of 'contact bounce'.

Better practice, virtually eliminating this disadvantage, is to use a bistable circuit based on the use of two 2-input NAND gates (Fig. 7.7). The mechanical switch S now simply earths *A1* in one position and *B2* in the other. When the switch is in position *A1*, the outputs at *Y1* and *Y2* are 1 and 0 respectively. When the switch is moved to leave *A1* floating and *B2* earthed, the output from gate 2 switches to logical 1 whereas the output from gate 1 switches to 0. Returning the switch S to *A1* restores the circuit to its original state as shown in Fig. 7.7.

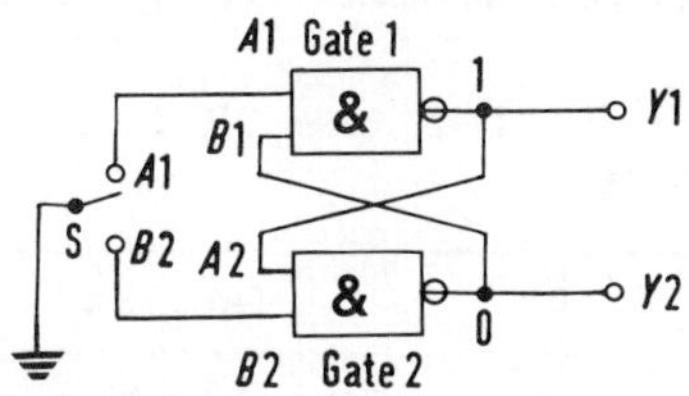

Fig. 7.7 A bistable circuit based on the use of two 2-input NAND gates to eliminate contact bounce.

In an experiment based on the circuit of Fig. 7.7, observe the output voltage levels on a cro. As the earth contact is moved from *A1* to *B2* and back, note that the transition is clean, free from any contact bounce.

The light-emitting diode circuit (7.5) can also be used to examine the behaviour of this bistable circuit.

7.4 *Further pulse generator circuits based on NAND gates*

Apart from the bistable circuit (Section 7.3.1) which makes use of two of these gates, the SN 7400N (Section 7.2.1) can be used to construct a free-running multivibrator and so provide a useful square-wave generator of which the output voltage switches between 0 and +5 V (valuable in a logic laboratory). In this circuit all four of the 2-input NAND gates of the SN7400N unit are interconnected. It also provides a monostable circuit, involving three of the 2-input NAND gates, acts as a source of single pulses of adjustable duration, and as a Schmitt trigger circuit (involving two of the 2-input NAND gates) which serves to shape poorly defined voltage signals of sufficient amplitude. Such shaping is usually needed for feeding such signals into a digital system. These further circuits are described in detail in Chapter 8.

7.5 *The OR and the exclusive-OR functions*

The OR and the exclusive-OR functions are compared by their respective truth tables (Table 7.2).

Table 7.2 Truth table for the OR function and for the exclusive-OR function

Input		Output	
A	*B*	OR	Exclusive-OR
0	0	0	0
0	1	1	1
1	0	1	1
1	1	1	0

The exclusive-OR logic gate - which is widely used - is also called the non-equivalent gate because the output is logical 1 when the inputs are not equal ie *A* = 1 and *B* = 0 and vice versa.

The use of four 2-input NAND gates (Fig. 7.8) enables the exclusive-OR function to be implemented. The LED circuit of Fig. 7.5 is used to determine the state of the

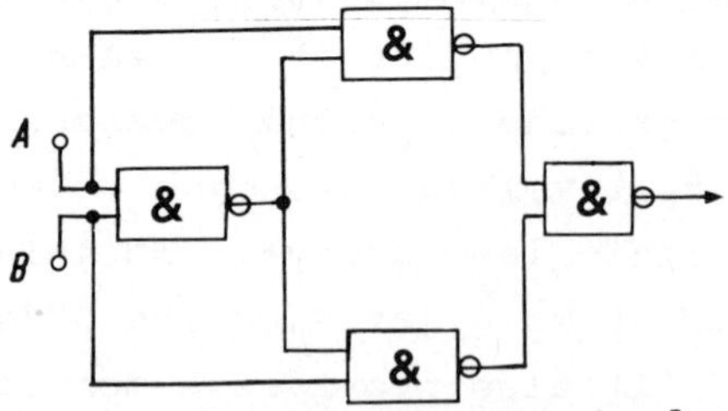

Fig. 7.8 An exclusive OR gate implemented by the use of four 2-input NAND gates.

output for various input conditions to the circuit of Fig. 7.8 and the exclusive-OR behaviour as shown in Table 7.2 is verified by experiment.

7.6 *Complementary metal-oxide semiconductor (CMOS) logic gates: the inverter (NOT) pair, the NOR gate and the NAND gate*

The basic element common to CMOS gates is the P and N 'inverter' pair shown with its equivalent circuit in Fig. 7.9. Consider the circuit of Fig. 7.9a: its inverter (NOT) function is clear because when a positive voltage ($+V_{DD}$) is applied to the input at A the p-channel device P turns off and the n-channel device N is turned on so that the output is equal to $-V_{SS}$: the terminal 'out' of Fig. 7.9a is joined to $-V_{SS}$ by the ON resistance (R_N of Fig. 7.9b) of the n-channel insulated gate fet (igfet). The action resulting when the input is negative ($-V_{SS}$) is obvious from Fig. 7.9b with the output equal to $+V_{DD}$ via the ON-resistance (R_P) of the p-channel insulated gate fet. The input resistance of the inverter pair is exceptionally large (typically 100 MΩ) whereas the output resistance is relatively small (typically 1 kΩ) as it is the series channel resistance of the conducting igfet. Thus the output signal voltage swings sharply between $+V_{DD}$ and $-V_{SS}$

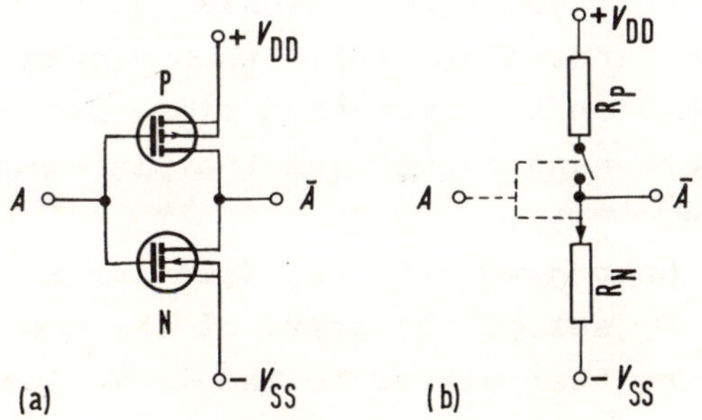

Fig. 7.9 (a) The CMOS inverter pair.
(b) The equivalent circuit.

Whereas TTL gates consume power during both the ON and OFF states, most of the power dissipated in a CMOS gate is due to the charging of the circuit capacitance and not (as one might expect) due to the current flow through the resistance of the conducting channel. This fact is confirmed on observing the rise of power consumption with an increase in switching speed.

The NOR and NAND functions are formed respectively by series and parallel combinations of the basic inverter pair (Fig. 7.10). Because a NOR gate with positive logic becomes a NAND gate with negative logic, the CMOS NOR gate is easily converted to a NAND gate simply by interchanging the p-and n-channel transistors.

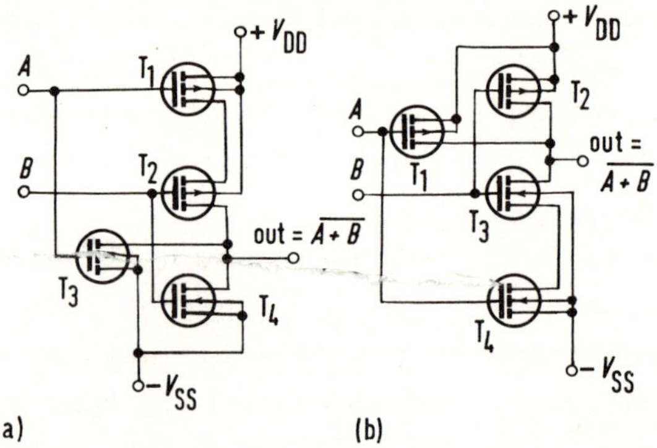

Fig. 7.10 CMOS gates: (a) NOR; (b) NAND.

7.6.1 *A NAND gate.* The pin arrangement of the CMOS dual-in-line 4011 package is similar to that of other such packages (including TTL). When viewed from above with the recess on the left-hand side, pin number 1 is in the bottom left-hand corner and the pin numbers increase anticlockwise.

Using one NAND gate with V_{DD} (pin 14) at +5 V and V_{SS} (pin 7) at 0 V, select the state of the two inputs A and B by connecting them either to the +5 V line (high) or the 0 V line (low). Using the logic probe (Section 7.1.3) examine the state of the output for the four possible configurations of the input.

Establish that the behaviour of the NAND gate is in

accordance with the appropriate truth table (Table 7.1). Note that with input A held at logical 1 the NAND gate operates as a simple inverter as it does when the two inputs A and B are linked.

With V_{DD} = +5 V use a DVM to measure the output voltage in the low and in the high state. Repeat these measurements with V_{DD} = +15 V. In each case, the output voltage swings from near zero to nearly $+V_{DD}$.

7.6.2 *The power consumption of a CMOS gate.* This is desirably very small but is dependent on (i) the supply voltage, (ii) the magnitude of the capacitance load and (iii) the operating frequency, ie the rate at which the gate is switched.

Set up the circuit of Fig. 7.11 in which each output is connected by means of a 15 pF capacitance to earth.

With a microammeter (μA) having a fsd of 100 μA, establish that the current flow from the power supply is too small to measure whether the outputs are low or high.

With V_{DD} = +5 V and a pulse generator providing pulses of height +4 V and the mark-to-space ratio unity (ie pulse duration equals half the repetition time) measure the average current demanded by the circuit when operating at 1 kHz, 10 kHz, 100 kHz and 1 MHz. Repeat this procedure with V_{DD} = +10 V and pulses of height 9 V.

Use the average current recorded with V_{DD} = +10 V and pulses of height +9 V at a frequency of 1 MHz to calculate the approximate power consumption of one NAND gate.

7.6.3 *The current flow to a CMOS gate in the ambient and switching modes.* Whereas in the experiment of Section 7.6.2 the microammeter (μA) records an approximate mean current to the circuit, the form of the current transients can be observed from the voltages developed across the 100 Ω resistor in the circuit of Fig. 7.11. To do this set V_{DD} at +15 V, the input pulse height at +14 V at a frequency of 300 kHz and display and record simultaneously on a double beam cro the voltage across the 100 Ω resistor and the output voltage from one NAND gate. These voltage waveforms should be as shown typically in Fig. 7.12. Note the current surge at the instant of switching and the negligible current flow through the gate in the ambient state.

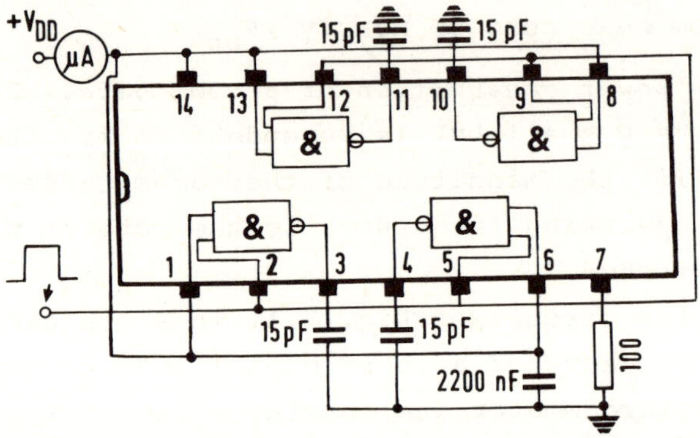

Fig. 7.11 The power consumption of a CMOS NAND gate. (An ic module 4011 is used).

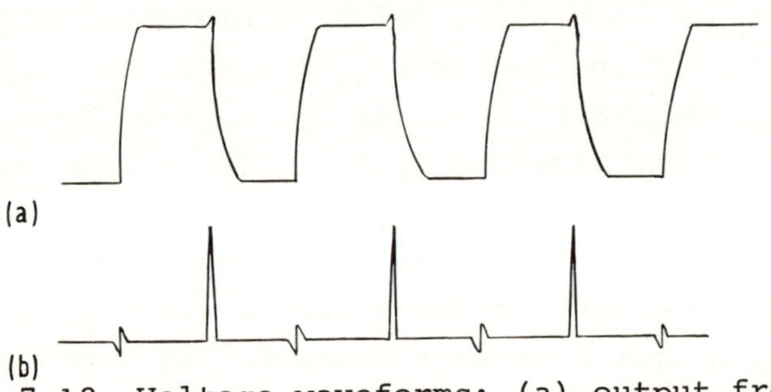

Fig. 7.12 Voltage waveforms: (a) output from one NAND gate; (b) across the 100 Ω resistor of Fig. 7.11.

7.6.4 *The power consumption of a TTL gate.* Making use of the 7400 TTL quad 2-input NAND gate, set up the circuit of Fig. 7.13. With V_{cc} = +5 V measure the current demanded by the circuit when all the outputs are low and when all the outputs are high. With a voltage pulse of height +4 V switch the gates at frequencies of 1 kHz,

10 kHz, 100 kHz and 1 MHz and, in each case, record the current using the milliammeter (mA) of Fig. 7.13. Also observe and record the voltage waveform across the 10 Ω resistor in the earth lead. As compared with CMOS, note the large current surge demanded in switching and the steady current flow in the circuit at all times.

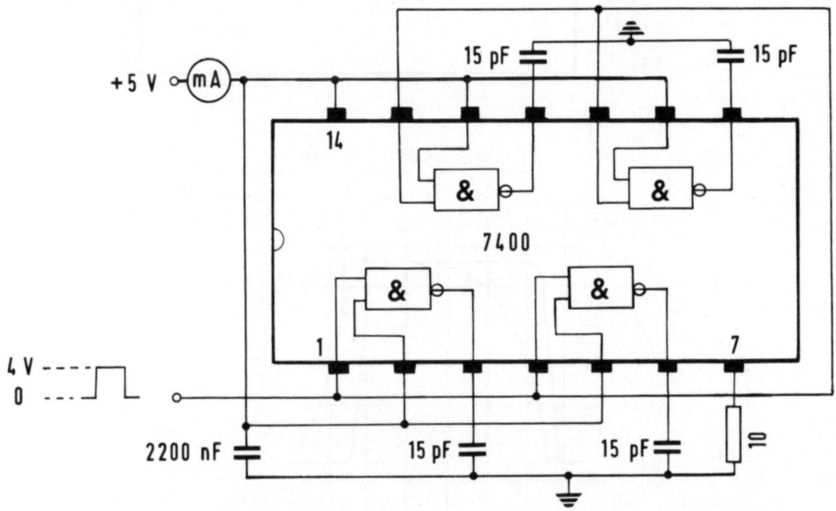

Fig. 7.13 The power consumption of a TTL gate.

Calculate the approximate power for one TTL NAND gate at a frequency of 1 MHz. Note that the current for this gate is virtually unchanged whether the switching is at 1 kHz or 1 MHz, unlike the CMOS gate.

7.6.5 *Rise time, fall time and propagation time for a CMOS gate.* The rise time (sometimes called the transition time from low to high and denoted by t_{TLH}) depends on the voltage supply and on the capacitative load. Connect together two of the NAND gates as in Fig. 7.14a, in which the capacitance load C_L consists of the capacitance of the coaxial lead to the cro plus the input capacitance of the cro itself. With input pulses of frequency 100 kHz and 0.5 μs duration, measure the rise times with the supply voltage V_{DD} at the values +5 V, +10 V and

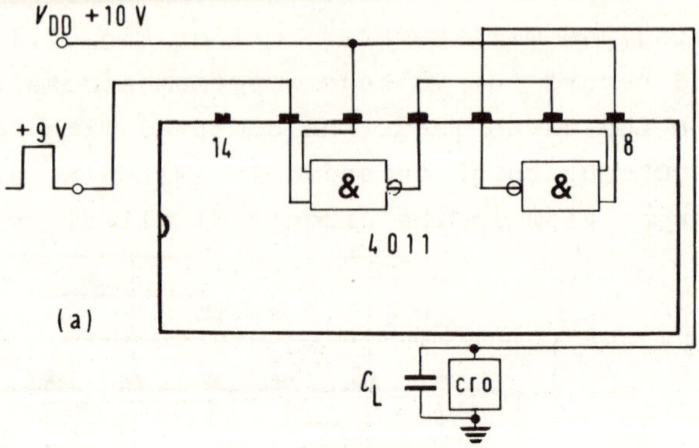

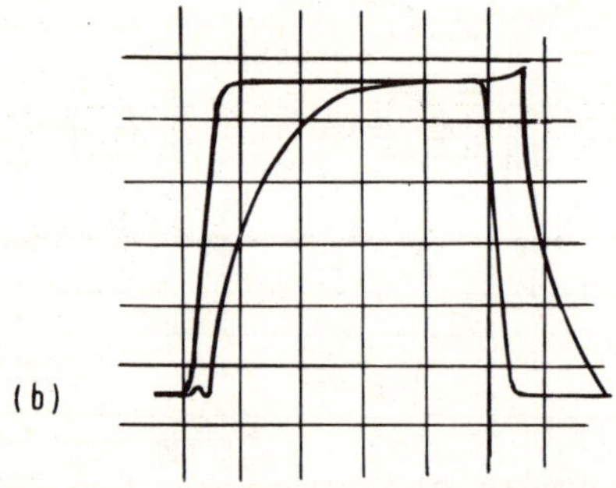

Fig. 7.14 (a) The effect of load capacitance and supply voltage on the rise time of a CMOS gate. (b) The voltage waveforms for the rise time, fall time and propagation time: 3 V/div, 0.1 μs/div 100 kHz input, pulse duration 0.5 μs. ($C_L = C_{external} + C_{cable} + C_{scope} =$ = 68pF + 29pF + 40pF = 137pF).

finally at +15 V. The pulse amplitude should be 1 V less than the prevailing value of V_{DD}. Show that the rise time decreases as the supply voltage is increased. Connect an additional capacitor of 68 pF between the input terminal of the cro and earth and establish that the rise time is increased as the capacitative load C_L is increased. Typical voltage waveforms are shown in Fig. 7.14b.

Show also (during these measurements) that the fall time (the transition time t_{THL} from high to low) is increased as C_L is increased and also that it decreases as

V_{DD} is increased.

7.6.6 *Further investigations.* With the circuit of 7.14a, vary the value of C_L and with successive values of V_{DD} equal to +5, +10 and +15 V measure (i) the rise time, (ii) the fall time, (iii) the propagation time delay LOW to HIGH and (iv) the propagation time HIGH to LOW. Typical results are given in Table 7.3.

Table 7.3 The effect of the supply voltage V_{DD} and the capacitance C_L on the rise time, fall time, propagation time of a CMOS gate.

C_L	5 10 15			5 10 15			5 10 15			5 10 15		
pF	Rise time ns			Fall time ns			Prop time ns			Prop time ns		
69*	275	145	125	125	70	60	80	40	40	95	60	45
84	320	170	145	140	80	65	83	45	42	98	65	50
102	380	195	170	160	85	70	85	50	45	100	66	52
137	475	250	205	195	100	85	88	54	47	105	68	53
170	585	300	255	225	120	95	90	55	50	110	70	55

column headings 5,10,15 are values of V_{DD} in volt.
*Capacitance of cro (including appropriate cable to it)

It is of interest to compare the rise and the fall times for a TTL gate which are each approximately 5 ns and where the propagation time is approximately 10 ns.

7.7 *Multivibrator circuits based on NAND gates of the CMOS type*

The various waveform generating circuits based on TTL NAND gates which are referred to in Section 7.3 can also be based on CMOS NAND gates. The appropriate CMOS unit is the 4011: it is a dual-in-line package similar in performance to the corresponding TTL unit, the SN7400N

(Section 7.2.1).

As in Section 7.3 and the immediately following subsections CMOS circuits of the free-running multivibrator, monostable and bistable types, can all be constructed. They are described in some detail in Chapter 8. Of particular interest is that a free-running multivibrator of very low frequency is readily based on CMOS gates because these gates have very high input impedances.

CHAPTER EIGHT

Some integrated circuits

8.1 *Introduction*

A wide variety (about 2000 different models) of integrated circuits (ics) are available commercially. From the constructional point of view there are two main kinds of monolithic integrated circuit: (a) the planar bipolar silicon transistor type and (b) those based on metal-oxide-semiconductor transistors (mosts or mosfets). Both kinds (a) and (b) are formed within the surface of a silicon chip (or slice) which has dimensions of typically 1.25 × 1.25 × 0.25 mm. The kind (a) comprises bipolar transistors and semiconductor diodes (if needed) formed by diffusion usually within an epitaxial surface layer of n-type or p-type silicon (not more than 12 μm thick) on the silicon chip together with the (usually vacuum-deposited) metallic interconnections and lands (terminals) necessary to form the circuit required. Type (b) involves a silicon oxide (SiO_2) film grown on the surface of a chip (usually of n-type silicon) and also contains the required (usually vacuum-deposited) metallic interconnections and lands. In the treatment of logic circuits (Chapter 7) an example of ics involving planar bipolar transistors is the TTL, whereas CMOS logic circuits exemplify the oxidized silicon chip integrated circuit.

Usually an ic is superior in performance, more reliable,

may be cheaper and is certainly more compact than the equivalent circuit constructed from discrete components. In fact, if an ic is available for the electronic operation required, it is usually good practice to use it.

Integrated circuits are many and varied so those mentioned below indicate some of their possibilities. Chosen here for experiments are (1) an electronic timer, (2) a voltage stabilizer, (3) a voltage-to-frequency converter and (4) waveform generators. Operational amplifiers, already considered in Chapter 6, are also usually in integrated circuit form.

8.1.1 *An integrated circuit electronic timer.* In scientific work it is often required to introduce a precise time delay between two electrically controlled events. Electronic timing circuits invariably make use of the charging and discharging of a high quality capacitor through a precision resistor, to provide time delays ranging from a few microseconds to several seconds. This type of delay circuit is constructed readily by making use of the 555, an ic timer (introduced in 1972 by the Signetics Corp.). The pin configurations of the T package and the 8-pin dual-in-line V package are shown in Fig. 8.1a.

Several other manufacturers now market 555 timers with identical pin configurations so that devices can be readily interchanged. Two separate 555 units (the 556 timer) can be obtained in a single encapsulation.

The operating principles of the 555 timer, used in the time delay or monostable mode, are described in relation to the equivalent circuit of Fig. 8.1b, in which the time interval is controlled precisely by two external components; a resistor R and a capacitor C. Whether or not the capacitor is allowed to charge is determined by the state of the bistable (flip-flop) circuit. In the ambient state pin 2 may be held at $+V_{CC}$, transistor T_1 is conducting and

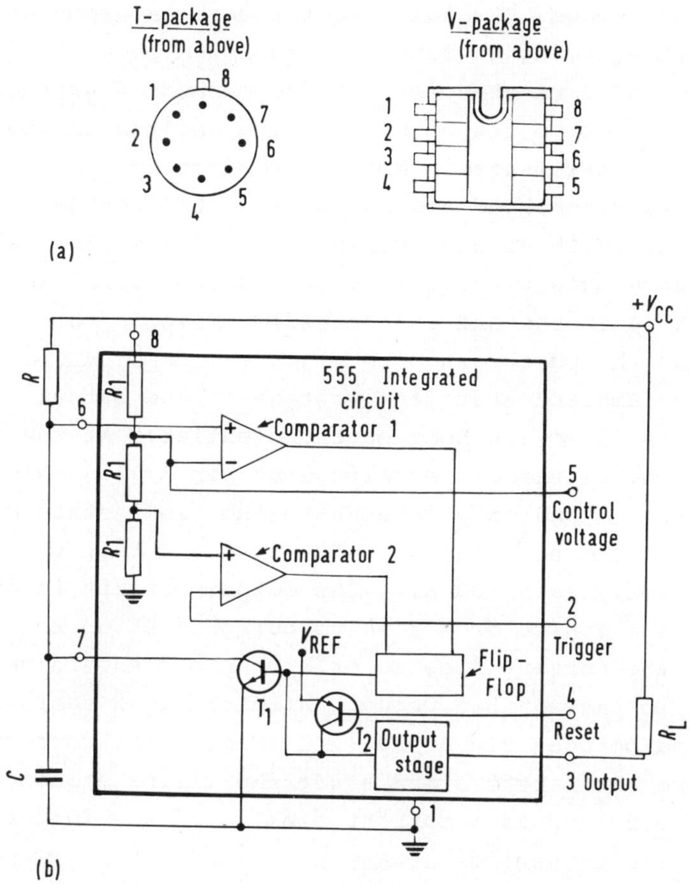

Fig. 8.1 The 555 electronic timer: (a) the pin configuration; (b) the equivalent circuit in the time delay mode.

the external timing capacitor *C* remains uncharged. A negative-going trigger pulse applied to pin 2 initiates the timing operation. The potential divider chain consisting of three equal resistors (each labelled R_1 in Fig. 8.1b) maintains the potential at the + input terminal comparator 2 at + $V_{CC}/3$. Hence the trigger pulse required to start the timing sequence must fall below $+V_{CC}/3$. When this trigger pulse is applied, comparator 2 switches the flip-flop which turns off T_1 and switches in the output stage. Capacitor *C* now charges exponentially through the timing resistor *R*. When the potential difference across *C* reaches $+2V_{CC}/3$, comparator 1 switches the flip-flop back to its original state and the timing interval is terminated. With T_1 conducting again, capacitor *C* is discharged and so is ready for the next timing operation. With no connection to pin 5 the time delay provided by this circuit is approximately 1.1*RC* seconds (where *R* is in ohm and *C* in farad). This is the time taken for the pd across *C* to reach $+2V_{CC}/3$. While the 555 is in its ambient state the voltage at the output terminal (pin 3) is 'low', ie near earth potential. At the start of the timing period the voltage at pin 3 goes 'high', ie close to $+V_{CC}$ and only returns to the 'low' state at the end of the timing interval. The rise and fall times are approximately each 100 ns. The output circuit is designed to act as a source or a sink to currents of up to 200 mA. so that the current flow to or from pin 3 should never be allowed to exceed this value. With the load resistance R_L connected between pin 3 and $+V_{CC}$ (Fig. 8.1b) current will flow through R_L before and after the timing sequence (the output is acting as a current sink). If the load resistance R_L is connected between pin 3 and earth, current flows through R_L from pin 3 only during the timing interval (the output is now operating as a current source). A miniature reed relay mounted in a dual-in-line package may serve as R_L so that power can be switched to other equipment for the selected time interval.

8.1.2 *Monostable or 'one-shot' operation of the electronic timer.* Set up the circuit of Fig. 8.2a with the supply voltage at +15 V, R = 10 MΩ, R_L = 500 Ω and C = 1 μF. In this ambient state the voltage at the output terminal (pin 3) is low, ie near earth potential, and current flows through the milliammeter (100 mA fsd) and R_L into pin 3: the output stage is acting as a current sink.

Initiate the timing sequence by pressing switch S to connect pin 2 to earth for an instant. The timing interval will be approximately $1.1 \times 10^7 \times 10^{-6} = 11$ s. Note that during this time the voltage at pin 3 is close to $+V_{CC}$ and the load current falls to a very small value. At the end of this interval of time the circuit reverts to its original state and appreciable load current flows again.

Repeat with R_L and the milliammeter connected between pin 3 and earth. Initiate the timing interval again and note that negligible load current flows before and after the time delay but that appreciable flow current flows for 11 s. The output stage is now acting as a source of current.

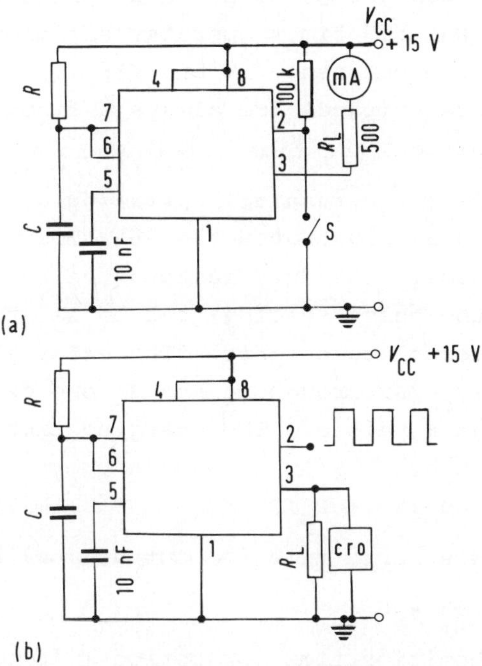

Fig. 8.2 Monostable operation of the electronic timer 555: (a) manual control; (b) pulse generator control.

Connect up the circuit of Fig. 8.2b with V_{CC} = +15 V, R = 10 kΩ, R_L = 10 kΩ and C = 100 nF. Use a pulse generator to provide pulses of amplitude +10 V, pulse duration 50 μs with a pulse repetition time of 100 μs to initiate the timing interval known to be approximately 1.1 ms. Display on a double-beam cro the voltages between pin 7 and earth and between pin 3 and earth. Vary R and C and provide experimental evidence to establish

(a) That the time interval is given by 1.1 RC (taking note of the tolerance in the stated value of any component)

(b) That the timing sequence is initiated when the voltage at pin 2 falls to $+V_{CC}/3$ and ends when the voltage across the capacitor reaches $+2V_{CC}/3$.

(c) That once the timing sequence is established, a subsequent negative-going pulse at pin 2 has no effect.

(d) That the 555 timer operates efficiently with V_{CC} set at any value between + 5 V and +15 V. As the comparator reference voltages are always a fixed fraction of the supply voltage, the time interval is independent of V_{CC}.

8.1.3 *Astable (free-running) operation of the electronic timer*. Fig. 8.3 illustrates the 555 timer connected in the astable mode. The duty factor (the ratio of the pulse duration to the pulse spacing) can be set by selecting the values of external components. The external capacitor C charges to $+2V_{CC}/3$ through R_A and R_B and discharges to $+V_{CC}/3$ through R_B alone. The charging time t_1 (to output *high*) is given by

$$t_1 = 0.69\ (R_A + R_B)C \tag{8.1}$$

whereas the discharge time (to output *low*) is t_2 given by

$$t_2 = 0.69\ R_BC \tag{8.2}$$

The pulse repetition time or period T is thus given by

$$T = t_1 + t_2 = 0.69\ (R_A + 2R_B)C$$

and the frequency by

$$f = 1/T = 1.44/|(R_A + 2R_B)C|$$

Set up the circuit of 8.3a with V_{CC} = +15 V, R_A = 5 kΩ,

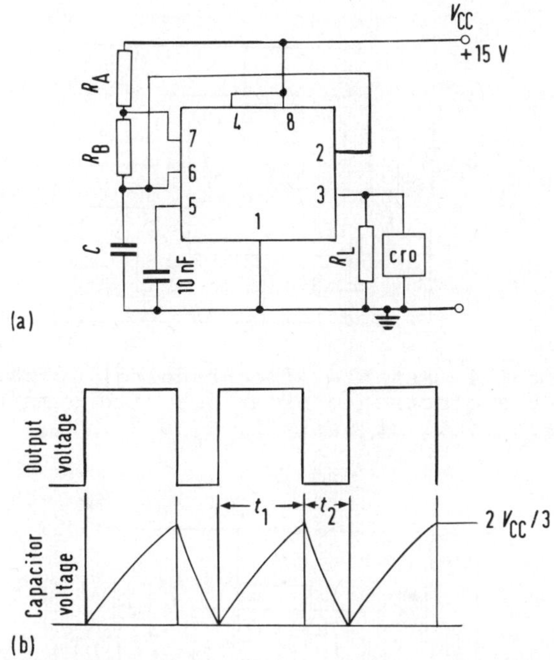

Fig. 8.3 Astable (free-running) operation of the electronic timer 555; (a) circuit arrangement; (b) waveforms.

R_B = 5 kΩ, R_L = 10 kΩ and C = 10 nF. Display and record on a double-beam cro the voltage waveforms at pins 6 and 3 (Fig. 8.3b is typical). Use the time-base of the cro to measure t_1 and t_2. Vary R_A and R_B (keeping $R_A \geq 1$ kΩ) and confirm the relationships (8.1) and (8.2) for t_1 and t_2 respectively.

In astable mode operation of the 555 timer, a mark-to-space ratio of unity can be provided if the external circuit is arranged as shown in Fig. 8.4. Examine the behaviour of this circuit with V_{CC} = +15 V, R_A = 5 kΩ, R_B = 5 kΩ, R_L = 10 kΩ and C = 100 nF.

8.1.4 *Use of the electronic timer (555) to provide a linear voltage ramp.* The external resistor R in the monostable arrangement (Section 8.1.2) is replaced by a constant current source so that the voltage across the

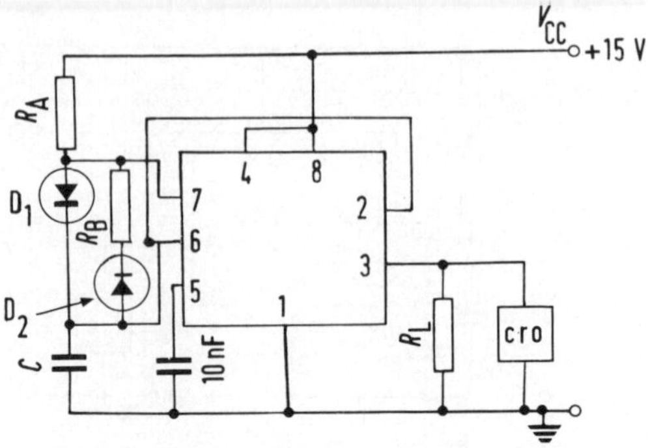

Fig. 8.4 Astable (free-running) operation of the electronic timer 555 with a mark-to-space ratio of unity if $R_A = R_B$.

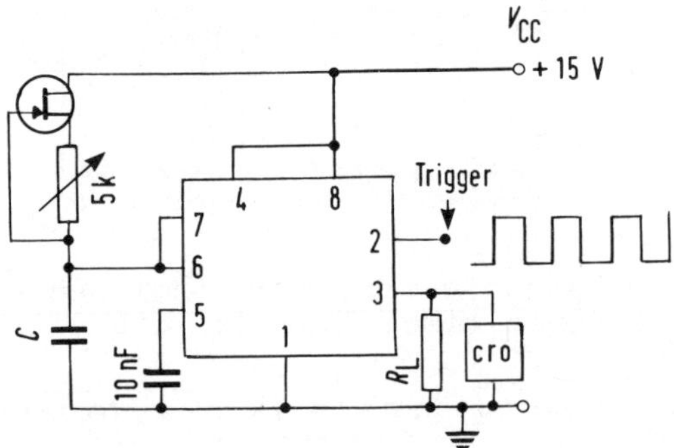

Fig. 8.5 A linear voltage ramp generator in which use is made of a constant current source with the electronic timer 555.

capacitor increases linearly with time from zero to $2V_{CC}/3$ The duration of the linear ramp is given by

$$t = 2V_{CC}C/3I \ .$$

Set up the circuit of Fig. 8.5 with V_{CC} = 15 V, R_L = 10 kΩ and C = 10 nF and initiate the timing sequence with a puls generator. Display and record simultaneously on a double-beam cro the voltage waveforms at pins 7 and 3. Record the appropriate cro amplifier and time-base settings so that quantitative results can be obtained.

8.1.5 *Effect of the control voltage on the monostable operation of the electronic timer 555.* In the experiments described so far the potential at pin 5 has been maintained at 0.67 V_{CC}. If this potential is set at some other value, preferably between 0.4 V_{CC} and 0.8 V_{CC}, the reference voltage on comparator 1 (Fig. 8.1) is altered and the time interval in monostable operation will no longer be 1.1 RC. Because each of the three resistors labelled R_1 in Fig. 8.1 has a value of about 5 kΩ, a convenient way of altering the potential at pin 5 is to connect a resistor R_2 (in excess of 5 kΩ) between pin 5 and $+V_{CC}$ or between pin 5 and earth.

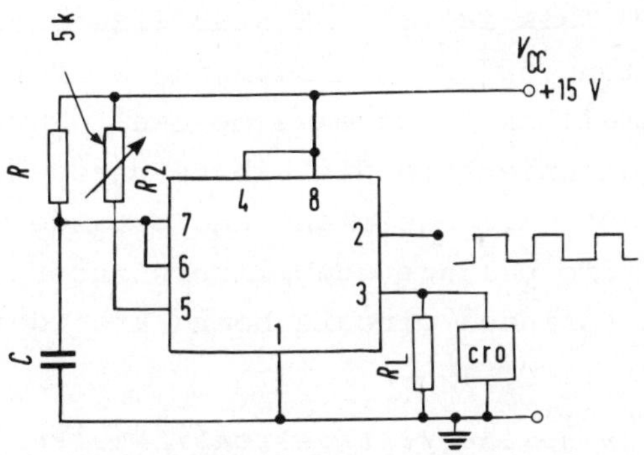

Fig. 8.6 Controlling the time interval of the electronic timer 555 by alteration of the reference voltage.

Connect up the circuit of Fig. 8.6 with V_{CC} = +15 V, R = 5 kΩ, R_L = 10 kΩ, $R_2 \geq$ 5 kΩ and C = 100 nF. Measure the potential (say, V_5) at pin 5 and adjust R (keeping its value above 5 kΩ) until V_5/V_{CC} has its maximum value. Measure the time interval under these conditions. Repeat with V_5/V_{CC} = 0.7. Now connect R_2 between pin 5 and the earth line. Vary V_5/V_{CC} between 0.4 and 0.66 (keeping $R_2 \geq$ 5 kΩ) and measure the time interval produced by the circuit for each value of V_5/V_{CC} .

8.1.6 *Further investigation.* The 'reset' facility available at pin 4 of the 555 timer has not been investigated. As long as the voltage at this reset pin is 'low' (near or at earth potential) the capacitor discharge resistor is ON and the capacitor cannot charge. If the voltage at pin 4 is switched low during a timing sequence the capacitor is discharged so the timing cycle is interrupted. The reset pin should be connected to $+V_{CC}$ when not in use. Investigate the use of this reset facility.

8.2 *A monolithic integrated circuit voltage stabilizer*
The voltage stabilizing section of the circuit of Fig. 6.32 (ie to the right-hand side of the input terminals) can be fabricated on a single silicon chip. The integrated-circuit voltage regulator provided is known as model MVR 12V. This ic voltage stabilizer, mounted in a TO3 encapsulation, offers the advantages of low cost, high performance, small size and ease of use. Furthermore, it is sometimes convenient to distribute unregulated voltage supplies to electronic equipment and provide voltage regulation locally; ic voltage stabilizers incorporated within the appropriate printed circuit board are ideal for this purpose.

8.2.1 *To determine the voltage stabilization factor S_V of the ic voltage regulator.* Experiment 6.3.9 is repeated using the ic voltage regulator MVR12V (Fig. 8.7). Across the output terminals is connected a 250 Ω (max) rheostat set to half its maximum value in series with an ammeter (1A fsd) constituting the load resistance R_L. A dc power supply is connected across the input terminals CD of the ic voltage stabilizer module and the input voltage V_i adjusted to 24 V. With a current I_L of 100 mA through the load resistance R_L the voltage V_o across this load resistance is measured. This is fixed at a nominal 12 V for this unit. Keeping I_L constant at 100 mA, V_i is varied in steps between 24 V and 16 V and, at each step, the output voltage V_o is recorded. V_o is plotted against V_i. The gradient of this graph gives S_V the voltage stabilization factor. The experiment should be repeated

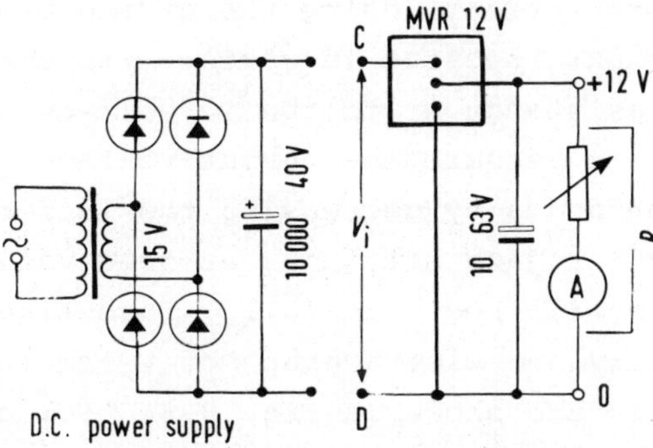

Fig. 8.7 Investigating the performance of the monolithic integrated circuit voltage regulator, MVR 12V (A is a 0-1 ammeter).

with I_L maintained at 200 mA.

To determine the output resistance R_o. Experiment 6.3.9 is repeated with the ic regulator MVR12V connected across the bridge rectifier and its smoothing capacitor. As no adjustment can be made to the output voltage of the ic regulator, the load current I_L should be increased in steps of 50 mA from 50 mA to 500 mA. At each setting the output voltage V_o is measured. From a plot of V_o against I_L gradient at I_L = 250 mA gives R_o in ohm.

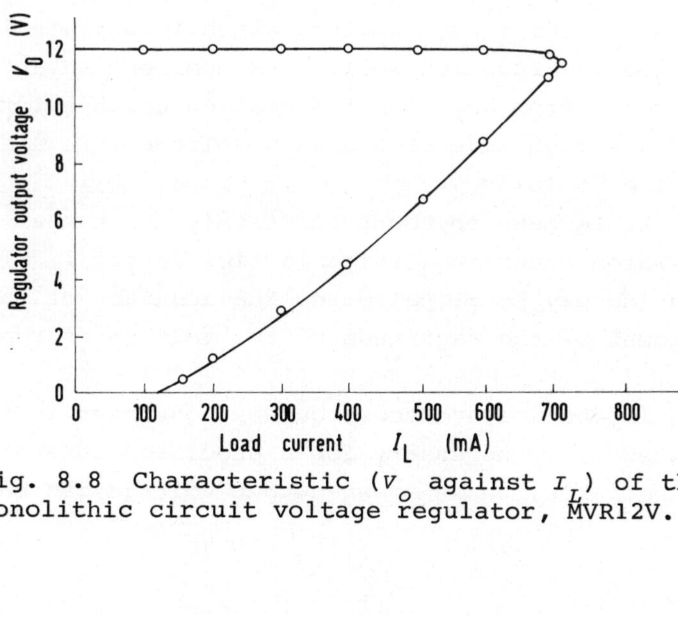

Fig. 8.8 Characteristic (V_o against I_L) of the monolithic circuit voltage regulator, MVR12V.

Overload protection. Using the arrangement of Fig. 8.7 with a 50 Ω (max) rheostat in series with a 0.1A ammeter as the load resistance R_L to the ic voltage regulator, with I_L = 250 mA measure the output voltage V_o. Increase I_L in steps of 50 mA by decreasing the load resistance and measure V_o at each setting. At some value of I_L the output voltage will begin to fall. Investigate this region carefully and plot a graph of V_o against I_L. Typical results are shown in Fig. 8.8. For obvious reasons this regulator is said to possess a fold-back overload characteristic.

8.3 *Voltage-to-frequency converters*

A voltage-to-frequency converter provides output pulses of a frequency (or pulse repetition rate prr) which is dependent on the magnitude of a steady voltage. Any voltage-controlled oscillator is able to do this, but of special interest are those converters for which the frequency is linearly related to the voltage. Several commercially available converters have a linearity of 0.01 percent and a temperature stability of 50 ppm per $^{\circ}C$ but they are expensive and are not considered further here. Instead a number of circuits are examined which illustrate important principles and function well. The manifest advantages which accrue from changing information which exists in the form of the magnitude of a steady voltage into a series of pulses are as follows:

(a) It is easy to count accurately the number of pulses which occur in a suitable time interval. The error in counting may be ±1 pulse, so the accuracy of the measurement of the magnitude of the voltage is readily found.

(b) If the voltage-to-frequency converter is to be calibrated it is necessary for a precision potentiometer. to be available, whose known output voltage can be varied.

Subsequent to such initial calibration, a voltage level can be measured rapidly and, if required, sampled several times a second.

(c) The voltage can be displayed in digital form, often (though not necessarily) preferable to reading a scale.

(d) In the form of pulses, a voltage signal can be transmitted over a considerable distance. Whereas voltage levels change appreciably along a conductor, it is not difficult to maintain the frequency and shape of transmitted pulses.

(e) A steady voltage level can be changed by an electrical disturbance ('noise'), whereas this noise need not affect a pulse repetition rate.

(f) A voltage-to-frequency converter can have a very high input resistance of 10 MΩ or more. In this respect, therefore, ideal voltmeter performance is approached.

8.3.1 *A voltage-to-frequency converter based on two operational amplifiers in an integrated circuit module.* A useful ic for experimental study is the 747: it contains two completely separate operational amplifiers (it is a dual opamp ic). Its pin configuration is shown in Fig. 8.9; it utilizes the 14-pin version of the TO116.

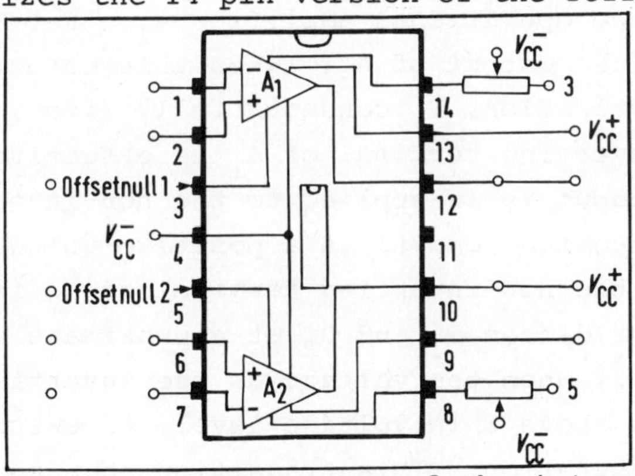

Fig. 8.9 Pin configuration of the integrated circuit module 747: a dual opamp.

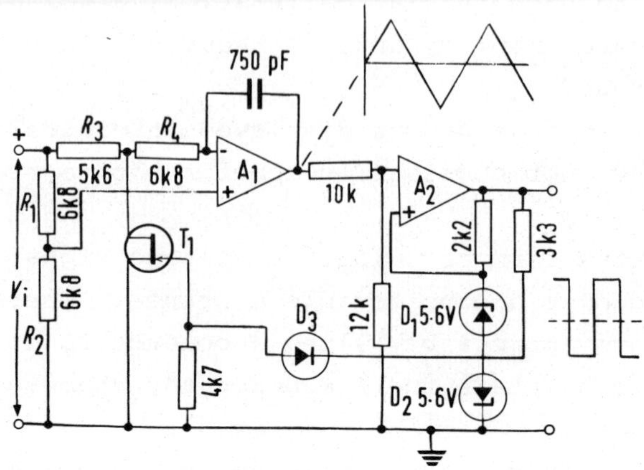

Fig. 8.10 Circuit arrangement of the voltage-to-frequency converter based on the 747.

Fig. 8.10 shows the circuit of a voltage-to-frequency converter based on these two operational amplifiers A_1 and A_2 where A_1 forms the basis of an integrator which provides a linear voltage ramp, whereas A_2 forms the basis of a voltage comparator. T_1 is a fet switch. This is switched from a conducting to a non-conducting (blocking) state and causes the ramp voltage from A_1 to change direction.

The input voltage V_i is applied simultaneously to both inputs of the operational amplifier A_1. Fet T_1 is controlled by the output of A_2. When this output of A_2 is at positive saturation, T_1 conducts fully (its V_{GS} = 0) so that the inverting terminal of A_1 is effectively earthed. Hence the input V_i is applied to the non-inverting terminal of A_1 which results in a positive-going output. Meanwhile, the non-inverting terminal of A_2 is maintained by the Zener diodes D_1 and D_2 at approximately +6.2 V (5.6 + 0.6). When the voltage at the inverting terminal of A_2 rises above this voltage level, A_2 switches to negative saturation. This transition changes the voltage at the non-inverting terminal of A_2 to -6.2 V and in addition turns T_1 hard off (gate negative with respect to source). The output of A_1 now begins to ramp in the negative direction and the cycle is repeated. Because the ramp rate is directly proportional to the input voltage V_i a linear voltage-to-frequency conversion is obtained.

In an experiment with this circuit the supply voltages to the operational amplifiers are set at ± 15 V. Then, by means of a potential divider network, a steady voltage of +1 V (measured with a dvm) is applied to the input terminal (ie $V_i = +1$ V) and the output waveforms from the integrator and from the comparator are displayed and recorded on a double-beam cro. The input voltage V_i is then raised in steps over the range + 50 mV to + 5 V and measured. At each known voltage setting, the output frequency is measured, preferably with a digital frequency meter. A graph of input voltage against frequency is plotted on log-log paper.

8.3.2 *An alternative voltage-to-frequency converter based on the 747 dual opamp ic.* In an alternative arrangement (Fig. 8.11a) to that of Fig. 8.10, amplifier A_1 acts as an integrator - and so forms the basis of a ramp generator - but amplifier A_2 now functions as a regenerative comparator. Initially assume that the output from A_2 is at positive saturation V^+_{osat}. Diode D_1 is reverse-biased and the output from the integrator falls linearly at a rate determined by the steady positive input voltage V_i. When the output voltage V_o from the integrator reaches a value given by

$$V_o = -(R_1/R_2)V^+_{osat}$$

the voltage at the point X (Fig. 8.11a) becomes zero so the output from A_2 switches to negative saturation V^-_{osat}, diode D_1 becomes forward biased and the integrator output ramps rapidly in a positive direction (Fig. 8.11b). Amplifier A_2 switches back to positive saturation when the

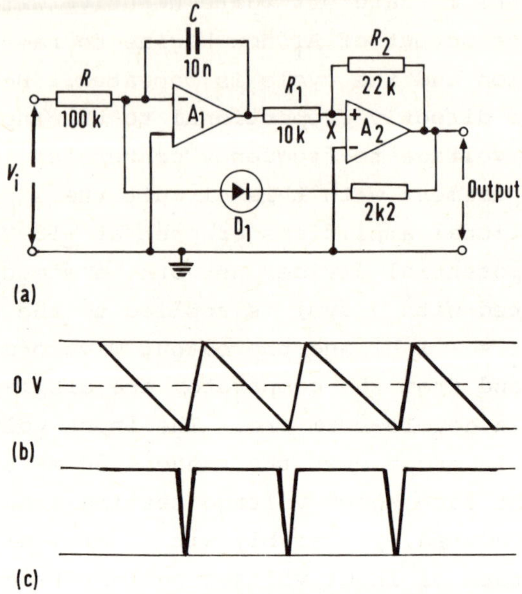

Fig. 8.11 An alternative voltage-to-frequency converter which makes use of a dual opamp such as the 747; (a) the circuit arrangement; (b) the output voltage waveform from the integrator based on A_1; (c) the waveform of the output voltage from the comparator based on A_2.

output voltage V_o from the integrator reaches a positive value given by

$$V_o = \mp (R_1/R_2)V^-_{osat}$$

The rate of change of the negative ramp voltage appearing at the output of A_1 is directly proportional to V_1. If the time taken by the positive ramp is negligibly small, a linear voltage-to-frequency conversion is obtained.

8.4 *Monolithic integrated circuit waveform generators*

Monolithic ic generators capable of producing outputs of a variety of waveforms are available: an example is the 8038 (manufactured by Intersil Inc.) which is able to produce an output of sinusoidal, square or triangular waveform with a frequency (or pulse repetition rate) which can cover a wide range and which can be selected by the connection to the ic of external passive components of appropriate values. This ic is mounted in a 14-pin plastic dual-in-line package (TO 116) with the pin configuration and functions shown in Figs. 8.12a and b respectively.

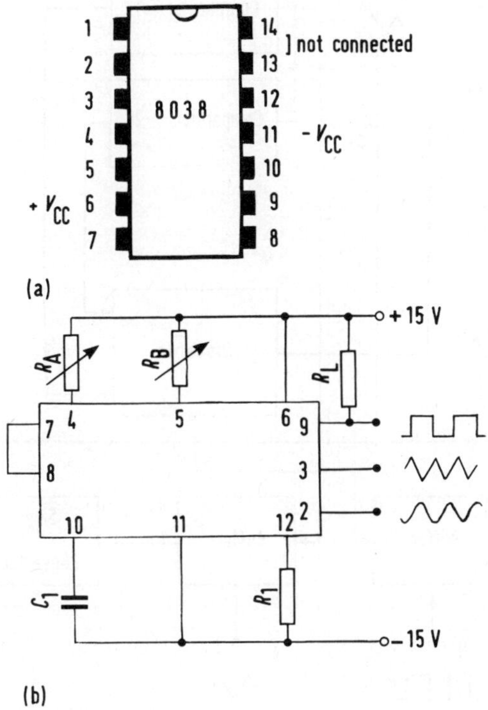

Fig. 8.12 The 8038 waveform generator: (a) pin configuration (a TO 116); (b) functions.

The operating principles of this waveform generator can be examined with the aid of the block diagram of Fig. 8.13. An external capacitor C_1 is charged and discharged by two constant current sources where current source 1 is on continuously and provides a constant current I whereas current source 2 is switched on and off by the bistable circuit and provides normally a constant current of $2I$ (I and $2I$ prevail if $R_A = R_B$ in Fig. 8.12b). When the current source 2 is switched off, the capacitor C_1 is

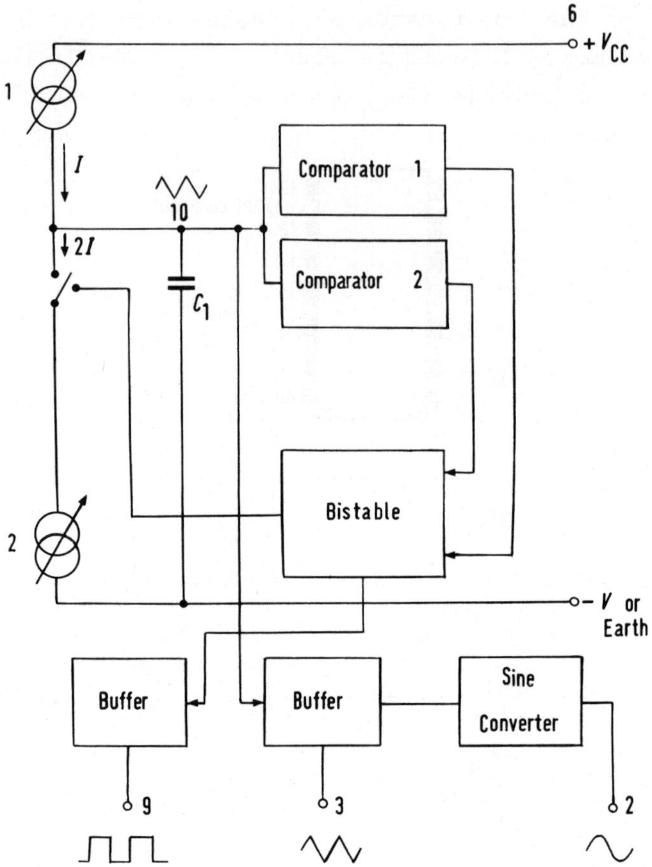

Fig. 8.13 Block diagram of waveform generator (type 8038).

charged in a linear manner from source 1 so the voltage across C_1 increased linearly with time. When this voltage reaches the level of comparator 1 (set at two-thirds of the supply voltage) the bistable circuit changes state and switches on the current source 2. As current source 2 usually provides a current $2I$, the capacitor C_1 is discharged by a net constant current I so the voltage across it decreases linearly with time. When this voltage reaches the level of the comparator 2 (set at one-third of the supply voltage) the bistable circuit reverts to its original state and the cycle is repeated.

The currents provided by the current sources can be altered by adjustment of the values of the external resistors R_A and R_B (Fig. 8.12b). With $R_A = R_B$ the source currents are I and $2I$ respectively, the charge and discharge times of the capacitor C_1 are equal, so a triangular voltage waveform appears across this capacitor. On the other hand, the bistable circuit produces a square wave (equal pulse durations and equal times between recurrent pulses).

If $R_A \neq R_B$ an asymmetrical sawtooth waveform is produced at terminal 3(Fig. 8.12a) and a rectangular waveform with a variable duty cycle at terminal 9 (Fig. 8.12a). The triangular waveform is shaped by a non-linear network to produce at terminal 2 an output of sinusoidal waveform.

Note that this ic module 8038 contains within a single small encapsulation several simple function generators, two constant current generators, two voltage comparators, a bistable circuit, a sine convertor and two buffer amplifiers.

Two points are of special interest:

(i) The device can feel quite hot to the touch when operating normally.

(ii) The device can be damaged if the output terminals are shortcircuited or if excessive current is demanded from the generator. Because no means of limiting the current exists at the outputs, the generator should operate into a circuit with an input resistance $\geq$ 15 kΩ. Alternatively, a simple voltage follower circuit can be connected to the output terminal in use: the 741 operational amplifier is useful for this.

8.4.1 *The power supply connections of the ic waveform generator.* This waveform generator (type 8038) can be operated from a single power supply (V = 10 V to 30 V) or a dual voltage supply (±5 V to ±15 V). With a single power supply the triangular and sinusoidal waveforms are symmetrical about a voltage of one-half the supply voltage, whereas the square waveform output voltage alternates between +*V* and earth. A dual power supply has the advantage that all the waveforms are symmetrical about earth.

Using a dual voltage supply (+15 V to 0 to - 15V)* connect up the circuit of Fig. 8.12b with $R_A = R_B = 100$ kΩ $R_1 = 81$ kΩ, $R_L = 15$ kΩ, $C_1 = 4700$ pF and pins 7 and 8 connected together. The mid-point of the power supply (labelled 0) should be connected to the common line. Display and record on a double-beam cro the triangular and the square waveforms. Note that the waveforms are symmetrical about earth when $R_A = R_B$. Measure the frequency. Note that the output from pin 9 is large while the capacitor is charging (ie during the positive slope of the triangular wave) and is small while the capacitor is discharging. Display on the double-beam cro the triangular and sinusoidal waveforms simultaneously: these must always be in phase (in that their peak values occur together) because one is derived from the other.

Reduce the supply voltages below ±15 V and record the lowest voltages at which the waveform generator still operates - this will be about ±5 V.

Disconnect the dual voltage supply from the module. Apply a single voltage of +30 V to pin 6 and earth pin 11 ie connect it to the earth terminal of the cro (Fig. 8.12a). Observe the output waveforms and measure the reference voltage levels about which the triangular and sinusoidal waveforms are symmetrical. Reduce the applied

*Some 8038 type integrated circuits have been found to overheat with a voltage supply of +15-0-15 V. It is hence advisable to operate at +12-0--12 V if the device is to operate for long periods.

voltage to below +30 V; operation should continue down to about +10 V.

The square wave output can be T TL compatible (ie compatible with transistor - transistor logic circuitry). To show this, connect the dual voltage supply (+15 V to 0 to -15 V) with the mid-point (0) to the common line and earthed via the cro. Disconnect R_L (Fig. 8.12b) from the +15 V line and connect it to the positive terminal of a separate 5 V supply of which the negative terminal is connected to the common line. The output from pin 9 is now a square wave which switches between +5V and earth.

8.4.2 *The use of external timing components with the ic waveform generator.* The symmetry of all the waveforms generated can be adjusted by variation of the values of the external timing resistors R_A and R_B (Fig. 8.12b) where R_A controls the rising portion of the triangular wave and of the sine wave and the low state of the square wave. The time t_1 for the rising portion of the triangular waveform is given by

$$t_1 = (5/3)\ R_A C_1 \tag{8.3}$$

whereas the time t_2 for the falling portion of this waveform is given by

$$t_2 = (5/3)\ R_A R_B C_1/(2R_A - R_B). \tag{8.4}$$

If $R_A = R_B = R$ it follows that $t_1 = t_2$ and a 50 percent duty cycle is achieved. The frequency f is given by

$$f = 1/(t_1 + t_2)$$

$$= 1/(5/3)\ \{R_A C_1\ |1 + R_B/(2R_A - R_B)|\ \}.$$

For $R_A = R_B = R$,

$$f = 0.3/RC_1. \tag{8.5}$$

Connect up the circuit (Fig. 8.12b) with $R_L = 15$ kΩ, $R_1 = 81$ kΩ, $C_1 = 4700$ pF and pins 7 and 8 connected together. Vary the timing resistors in steps over the range

$R_A = R_B = 100\ \text{k}\Omega$ to $R_A = R_B = 1\ \text{k}\Omega$, record the frequency f at each setting and note how the waveforms deteriorate at both ends of the frequency range. Present the results graphically to verify the form of Equation (8.5).

With $R_A = 22\ \text{k}\Omega$, vary R_B to alter the duty cycle. Use the time-base on the cro to measure the rise-time t_1 and the fall-time t_2 of the triangular waveform. Compare the measured values with those calculated from Equations (8.3) and (8.4).

8.4.3 *Independence of the frequency of the supply voltage for the ic waveform generator.* The switching actions occurring in the waveform generator are not dependent on the supply voltage even though these voltages are not regulated within the circuit. This is because both currents and thresholds are direct, linear functions of the supply voltage so that their effects cancel.

To investigate any variation of the frequency with the supply voltage, set up the circuit of Fig. 8.12b with $R_A = R_B = 7\ \text{k}\Omega$, $R_L = 15\ \text{k}\Omega$, $R_1 = 81\ \text{k}\Omega$ and $C_1 = 10$ nF and pins 7 and 8 linked. With the supply voltage at ±5 V (measured preferably with a DVM) record the frequency of the output on a digital frequency meter. Increase the supply voltage in steps of ±1 V up to ±15 V and record the frequency at each setting. Plot a graph of frequency against supply voltage. It should be borne in mind that the frequency is temperature dependent and that no attempt has been made to maintain the device at constant temperature in this experiment.

8.4.4 *Sine-wave distortion of the ic waveform generator.* This is inevitable to some extent because the sine wave is constructed from the triangular wave. To minimise this distortion, the 81 kΩ resistor used for R_1 is replaced by a variable 100 kΩ (max) resistor in series with a fixed 4.7 kΩ resistor and the effect on the sine wave of varying the 100 kΩ (max) resistor is observed. In this way a distortion of less than one per cent can be achieved. It is difficult to detect a small amount of sine wave distortion so it is usually convenient to compare the sinusoidal waveform with one of similar amplitude and frequency obtained from a good quality sinusoidal oscillator. Note whether the sinusoidal waveform is improved or otherwise as the frequency is increased.

8.4.5 *The ic waveform generator as a voltage-controlled oscillator.* The 8038 can be used as a voltage-controlled oscillator (vco) or a voltage-to-frequency converter. In the circuit (Fig. 8.14) $R_A = R_B = 3$ kΩ, $R_L = 15$ kΩ, $R_1 = 81$ kΩ and $C_1 = 10$ nF. Apply a voltage V_f (from a separate voltage supply) to this circuit between pins 8 and 6 so that pin 8 is negative with respect to pin 6. With $V_f = 10$ V, measure the frequency of the output between pin 9 and earth by means of a digital frequency meter; reduce V_f from 10 V to zero in steps of 1 V and measure the frequency at each setting. Plot a graph of frequency against voltage V; repeat with $C_1 = 100$ nF.

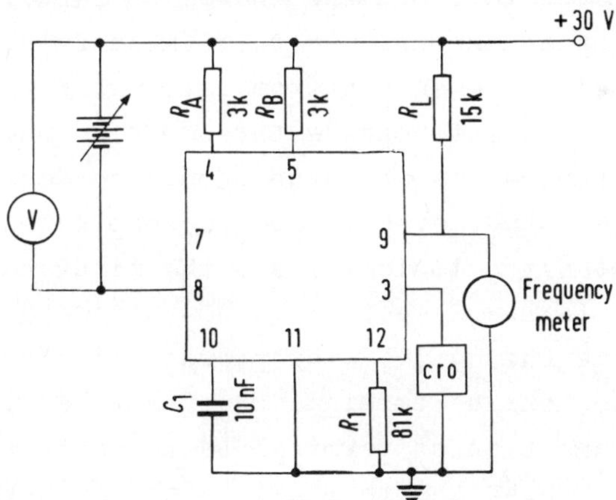

Fig. 8.14 The ic waveform generator (type 8038) as a voltage-controlled oscillator.

8.5 *Waveform generators of the multivibrator type based on NAND gates*

As indicated in Sections 7.4 and 7.7 multivibrator circuits of the free-running, monostable and bistable types can be based on TTL NAND gates as in the module SN7400N and also on CMOS NAND gates as in the module 4011.

8.5.1 *Astable (free-running) multivibrators based on (a) a SN7400N and (b) a CMOS unit 4011.* (a) A circuit (Fig. 8.15a) in which four 2-input NAND gates are interconnected (using the SN7400N unit) provides a square-wave output signal with switching between 0 and +5 V. Fig. 8.15b shows the appropriate ic module connections. Note that the NAND gates G1, G2 and G4 each have their two input terminals (A,B) linked so that each gate acts as an inverter (or NOT gate).

Consider the situation when the output from NAND gate *G1* switches from 1 to 0 (here 1 and 0 are 'logic levels', 1 usually corresponds to a voltage high enough to correspond to ON whereas 0 corresponds to a low voltage corresponding to OFF). Gate *G2* will invert this signal so that its output moves to 1 but the charge on capacitor C_1 cannot change instantaneously so that the terminal *A3* remains at the 0 level. The output from gate *G3* is therefore 1. Capacitor C_1 begins to charge through R_2. When the voltage on terminal *A3* is large enough to correspond to logical 1 the output of *G3* drops to zero. This switches the output of *G1* to logical 1 and the first half of the cycle is complete. Capacitor C_1 now discharges through R_2 to complete the cycle. The process continues indefinitely so long as terminal *B3* of gate *G3* is in open circuit. Connection of terminal *B3* to earth stops the oscillation. Thus, the terminal *B3* can be used as an ENABLE input to switch the oscillator ON or OFF. Gate *G4* serves as an inverter to provide a complementary output signal.

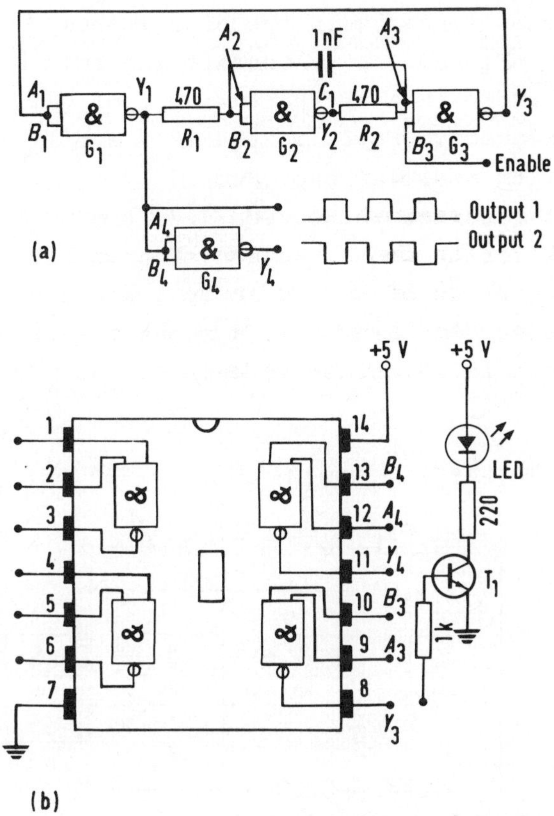

Fig. 8.15 A free-running multivibrator based on four 2-input NAND gates (in an SN 7400N module): (a) the circuit arrangement; (b) the appropriate ic module connections. T_1 is eg a BC109.

With $R_1 = R_2 = 470\ \Omega$ (as in Fig. 8.15) measure the mark-to-space ratio, ie the ratio of the pulse duration to the pulse interval. With the component values shown in Fig. 8.15 the pulse repetition rate or frequency is approximately 1 MHz.

The repetition rate is readily altered by changing the capacitance C_1.

Measure the frequency of the output from this multivibrator by means of a digital frequency meter. Display both output signal waveforms on a double-beam cro and record them. Repeat with the capacitance C_1 equal to 10 nF, 100 nF, and 1 μF. Confirm the role of the enable terminal. Note that the performance of this integrated circuit is much improved if the +5 V supply is decoupled to earth with a 0.1 μF capacitor.

(b) As asserted in Section 7.7, the CMOS NAND gate is attractive for the construction of an astable multivibrator which is of particularly low pulse repetition rate (frequency). Fig. 8.16 a shows the arrangement to achieve this in which use is made of the CMOS quad 2-input

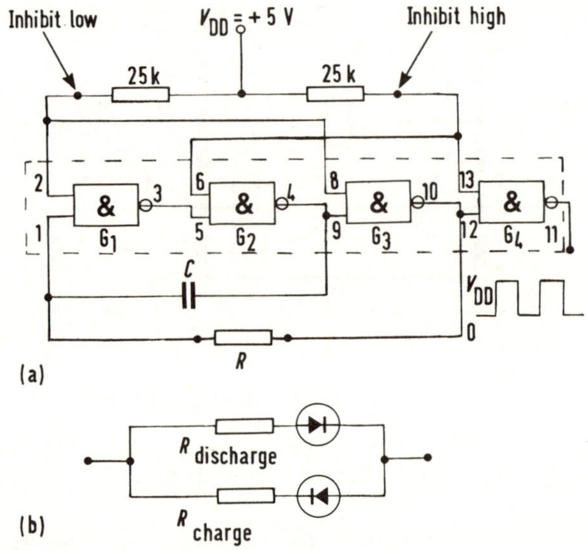

Fig. 8.16 (a) A low frequency multivibrator based on the CMOS unit 4011. (b) Circuit to replace R in (a) so as to permit the mark-to-space ratio to be varied.

NAND gate unit 4100. In this circuit the gates operate as simple inverters with the second inputs each used to provide an inhibit function. The inhibit pins are 2,6,8 and 13; they are connected via two resistors (each 25 kΩ) to the positive of the supply line at V_{DD} = +5 V, consequently they are at logical 1. The three gates G_1, G_2 and G_3 form the multivibrator, whilst the fourth gate G_4 acts as an output buffer. Gate G_1 monitors the potential at the junction of the capacitor C and the resistor R. When the potential at this junction falls to below the value of the threshold potential of gate G_1, one terminal of the capacitor C is connected to earth by gate G_2 whereas gate G_3 connects one end of the resistor R to $+V_{DD}$. The capacitor C hence charges up through R until the potential at the input of gate G_1 exceeds the threshold value. Then the output voltage of gate G_1 falls to zero, the output voltage of gate G_2 increases to $+V_{DD}$ and that of gate G_3 drops to zero. The result is that the gates have now connected the end of resistor R to earth and the capacitor to $+V_{DD}$, so the capacitor discharges through R until the gate G_1 again changes state. The circuit will therefore produce pulses at a repetition rate determined by the values of C and R with a mark-to-space ratio close to unity.

In an experiment, set up the circuit of Fig. 8.16a with C = 1 μF and R = 1 MΩ. Monitor the output voltage from gate G_4 with the logic probe (Section 7.1.3). Use a number of values of C and R and obtain a relationship between CR and the time period of the oscillator. The oscillations cease when one of the inhibit inputs is connected to earth. The output at pin 11 will be held in either a 1 or a 0 state depending on which inhibit input is earthed. Examine the effects of earthing each input in turn and ensure that the resulting behaviour of the circuit is in accordance with the truth table 7.1.

If the circuit of Fig. 8.16b is used to replace the resistor R in the circuit of 8.16a, the two diodes separate the paths of the charge and the discharge of the capacitor C. By selection of the values of the resistors ($R_{discharge}$ and R_{charge}) in each path, the mark-to-space ratio can be varied over a very wide range.

8.5.2 *An ic monostable multivibrator.* The SN74121 is suitable, it is contained in a plastic 14-pin dual-in-line (DIL) package of which the pin configuration is shown in Fig. 8.17.

Fig. 8.17 Pin configuration of the SN 74121 monostable module.

Set up the circuit shown in Fig. 8.18. With the switch S_1 open the input is high (ie at +5 V). If S_1 is closed the input is connected to earth and the negative input step initiates the transition. The output pulse width is given by $t = 0.69\ RC$ so that with the component values shown a pulse width of approximately 3 s should be obtained. A light-emitting diode (LED) in the output circuit enables examination of the state of the output and the duration of the output pulse.

To confirm that $t = 0.69RC$, choose suitable known values of R and C and initiate the transition with a pulse generator producing positive-going pulses occurring between 0 and + 4V. Display the input and output pulses on a cro. Note that the input voltage falling from +4 V to zero initiates the transition.

8.5.3 *A Schmitt trigger circuit based on two 2-input NAND gates.* In the circuit (Fig. 8.19) the two NAND gates can be selected from the four available in a SN7400N unit (Section 7.2.1); the Zener diode D_2 limits the input voltage to NAND gate G_1 to a safe level while the resistor

R_1 limits the current through D_2. The resistor R_2 provides the positive feedback (necessary to ensure fast switching) from the output of NAND gate 2 to the input to NAND gate 1.

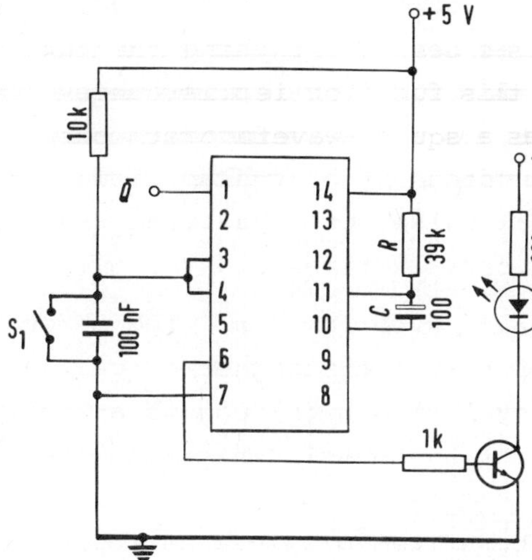

Fig. 8.18 Using the SN 74121 monostable module

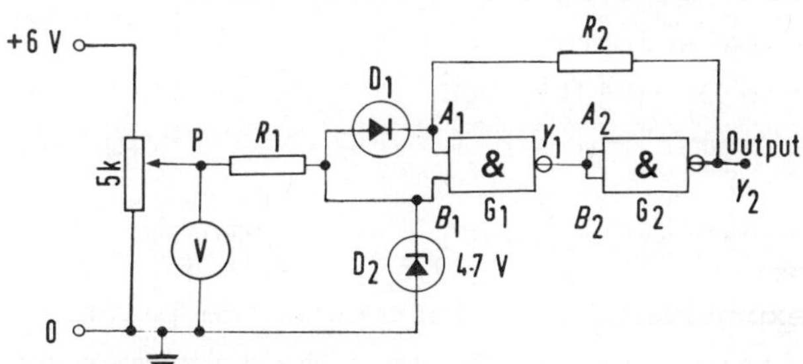

Fig. 8.19 A Schmitt trigger circuit based on two 2-input NAND gates.

Using a light-emitting diode (LED) to determine the state of the output, set up the circuit of Fig. 8.19. Measure the steady input voltages required for the transitions to occur. Remove the potentiometer providing the steady voltage to the point P and use a signal generator to provide a sinusoidal input signal of frequency 1 kHz and amplitude 10 V. Display and record the input and output waveforms with a double beam cro. Vary the amplitude of the input signal and note that it has no effect on the amplitude of the output; note that the frequencies of the input and output signals are the same.

The normal use of this type of circuit is to shape poorly defined pulses before using them as the inputs to a digital system: this function is achieved because the Schmitt circuit has a square-waveform output whose magnitude is independent of the amplitude and shape of the input signal pulse, provided that this pulse has a height above a certain threshold value.

8.5.4 *Additional investigation.* The three fundamental requirements of a Schmitt trigger circuit are (a) to act as a voltage level sensor (b) act as a regenerative switch (c) have inherent hysteresis which can preferably be varied.

As stated above its main use is for pulse shaping, ie to convert a positive-going voltage pulse with poor leading and trailing edges into a well-defined pulse with fast leading and trailing edges.

Make use of a 4011 CMOS ic to construct a Schmitt trigger circuit.

8.6 *A decade counter and a cold-cathode number display tube*

A set of experiments to introduce the principles utilized in electronic counters forms a good example of the use of the Texas Instruments 74 series of ic modules, ie a series of transistor-transistor logic (TTL) gates. In these inter-connected experiments, light-emitting diodes (LEDs) are used to indicate the numerical state of sections of the assembly of ic modules, but the total count is displayed by a gas-filled numerical indicator tube (number tube).

The input voltage pulses to be counted are from a switched battery supply (Fig. 8.20) and these are shaped

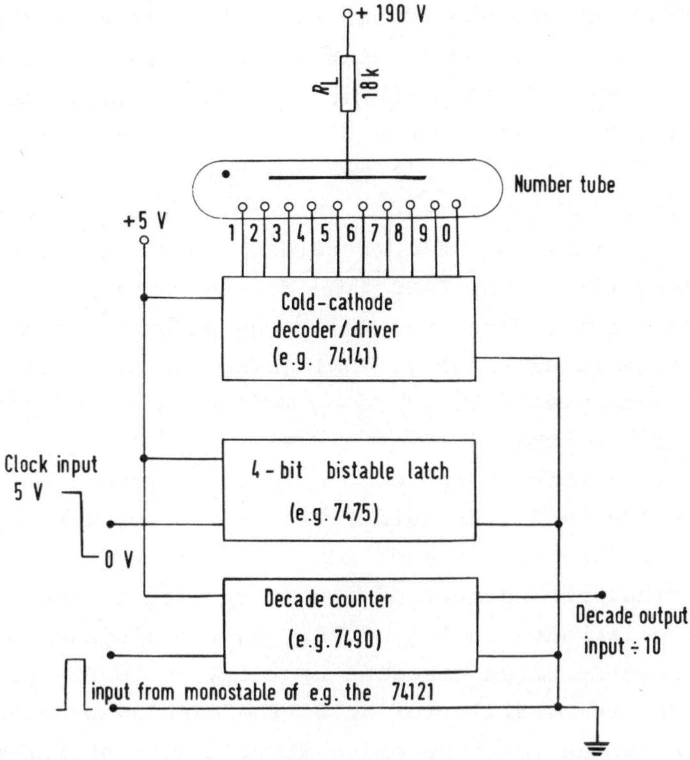

Fig. 8.20 The assembly of modules from the 74 series with a number tube output display, forming a decade counter.

by a monostable multivibrator.(the 74121 in the 74 series) and summed by the decade counter (the 7490 in the 74 series). The total count, at any instant, at the output of this decade counter is in binary coded decimal (BCD) form. This is to be displayed as a decimal number on the number tube so that conversion from data in BCD form to decimal form is needed; this is done by a decoder (the 74141 in the 74 series). As the outputs from this decoder are suitable to drive the gas-filled number tube (a XN12), this 74141 is called a decoder driver. In the assembly of units of the 74 series, a 4-bit bistable latch (the 7475 in the 74 series) is connected between the decoder/driver and the number tube; this enables the BCD output from the decade counter to pass directly to the decoder/driver if the state of the 'clock' input to this latch is logical 1. However, if the state of this clock input to the latch is switched to logical 0, the information which is contained at that instant is held in the latch. Likewise held (retained) is the display on the number tube. The 7490 (the decade counter) may continue to acquire data even while the output display on the number tube is stationary. Switching the clock input to logical 1 enables display to be made of any new state of the decade counter.

To effect this change of the logic state of the clock input to the 7475 (the latch) use is made of the toggle switch labelled 'HOLD' and 'RUN' shown in Fig. 8.21 which is the final arrangement of circuitry used in the experiments and is based on Fig. 8.20. As the display in the number tube is often changing rapidly, it is clearly convenient temporarily to 'hold' the display because this enables a value to be recorded at will without interrupting the count.

The voltage pulses to be counted are provided from one of the following: (a) a manually-operated press-button switch; (b) a telephone dial; (c) a low-frequency multivibrator. In the experiments, the counting set-up is operated very slowly; in fact, it can be operated at 18 million times per second (ie at frequencies up to 18 MHz).

Note regarding Section 8.6. The cold-cathode number display tube is being replaced in commercial instruments by the seven-segment light-emitting diode (led) display. However, despite the obvious disadvantage that the cold-cathode tube requires a supply of +190 V in a small digit digital system operating from a +5 V supply, these tubes provide a large display easily visible at some distance.

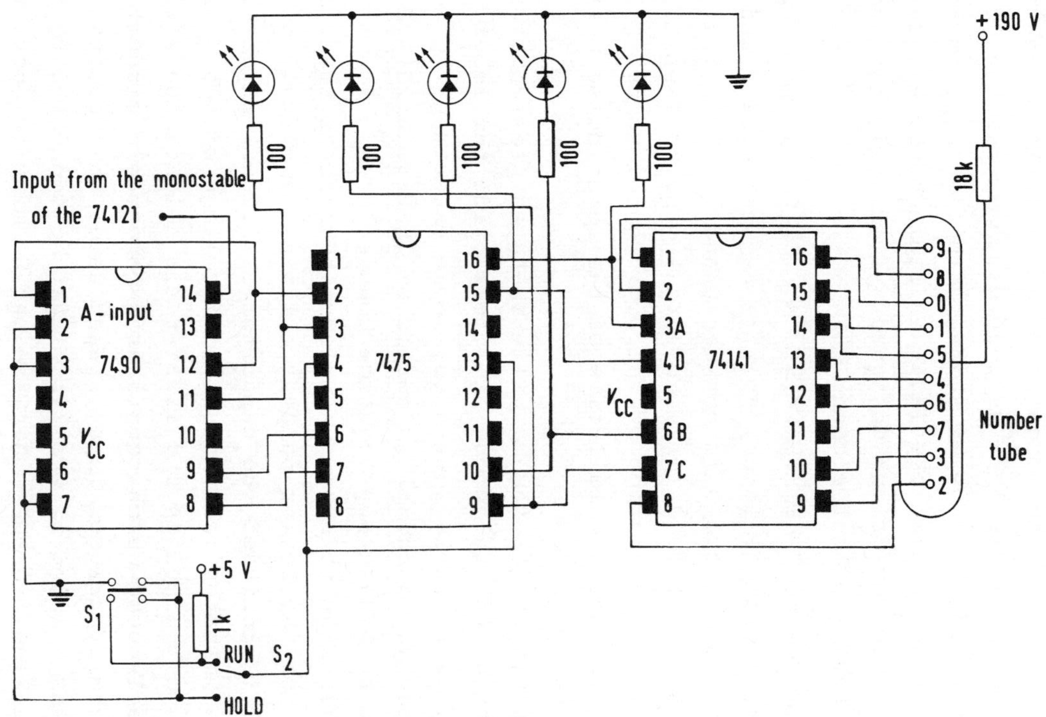

Fig. 8.21 The decade counter circuit.

Gas discharge tubes of this cold-cathode type for display are becoming more difficult to obtain and now they are relatively expensive. Nevertheless, the operating principles of the counting stage have been established in section 8.6 and it is left as a development exercise to replace the tube by a seven segment display based on leds. This will require that the 74141 is replaced by a 7447 which is a decoder/driver designed specifically for use with the seven segment element.

8.6.1 *Shaping the input pulses*. For positive logic with TTL gates, logical 1 is any voltage between +2.4 V and +5 V, whereas logical 0 is between 0 and +0.4 V. The transition from 1 to 0 (or vice-versa) must be fast to ensure reliable operation of the logic gate, so neither the rise-time nor the fall-time of the input pulse must exceed 1 μs.

Making use of (a) the manually-operated press-button switch or (b) the telephone dial to provide the input voltage pulses, it is convenient in an experiment to connect these two (a and b) in series as in Fig. 8.22. It is necessary to reduce any 'contact bounce' which gives rise to spurious pulses and also to shape the pulses. For the former purpose, within the circuit of Fig. 8.22 is included the capacitor C (2.2 μF) which largely eliminates 'contact-bounce' voltage spikes, whereas to shape the pulses and limit their heights the 74121 ic module is used which comprises a monostable circuit with a Schmitt trigger input. When using the Schmitt input, pins 3 and 4 of the 74121 <u>must</u> be connected to the zero volt line (earth) otherwise the Schmitt trigger input is eliminated. In a first experiment, therefore, the TTL input terminals (Fig. 8.22) are electrically connected.

Referring mainly to Fig. 8.21, after connecting the appropriate voltage supplies, plugging in the telephone dial and linking the TTL input to the zero-volt line

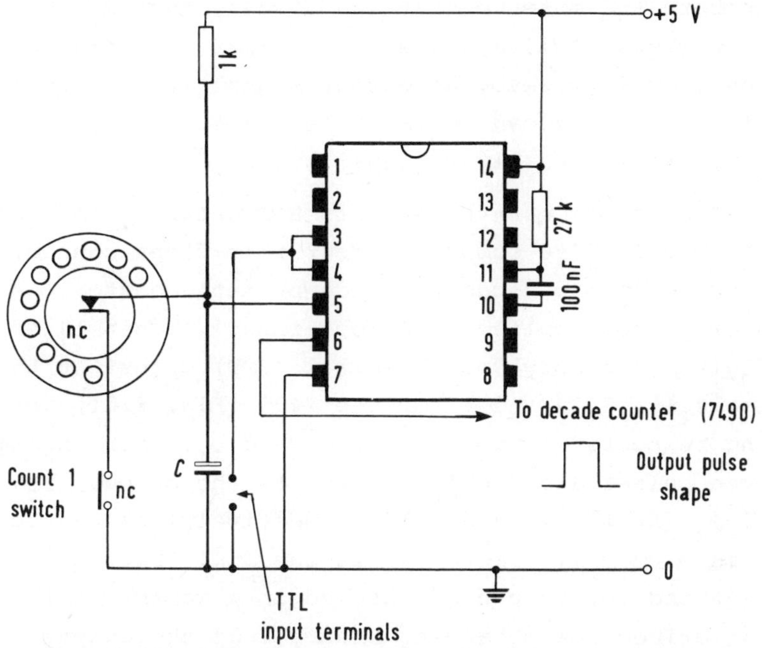

Fig. 8.22 The input pulse shaping circuit

(Fig. 8.22) the switch S_2 is set at RUN. The numerical indicator (number) tube can be set to zero by depressing the RESET switch S_1. On closing the 'count 1' switch (Fig. 8.22) an input pulse is applied (note that the telephone dial switch is normally closed, nc) and the number tube read-out should change from 0 to 1. Check that the number displayed is increased by unity each time the 'count 1' switch is closed (note that this switch is also normally closed).

Investigate the use of the 'HOLD' and 'RESET' facilities.

8.6.2 *Dialling-in decimal numbers to be added.* Remove the back of the telephone dial and confirm that the contacts are normally closed (nc). Dial, for example, the number 5 and note that the contacts open fives times. Do not touch the contacts and after this preliminary inspection replace the back-plate so that dust is excluded. Dial few numbers and observe that the display records the sum of the numbers dialled. When the number tube display changes from 9 to zero, an output signal (the carry signal) from the decade counter is available to trigger the next stage, usually a 'tens' stage.

8.6.3 *The decade counter and its BCD output.* The decade counter module 7490 comprises four flip-flops connected together to form a decade (divide-by-ten) counter. A train of pulses applied to 'clock-input' A (pin 14 of the 7490) gives a binary coded decimal (BCD) output at the terminals 11, 8, 9 and 12 of the 7490 (Fig. 8.21) corresponding respectively to Q_D, Q_C, Q_B and Q_A. For example, if seven pulses are applied to pin 14 the outputs Q_D, Q_C, Q_B and Q_A (at 11, 8, 9 and 12 respectively) should be 0111 which is binary 7. Table 8.1 is the full truth table for the 7490 module; it can be checked in a manner similar to that described for obtaining binary 7 at the output terminals 11, 8, 9 and 12.

Table 8.1 Truth Table for the 7490 (a TTL decade counter module).

Decimal number (count)	Output BCD			
	Q_D	Q_C	Q_B	Q_A
0	0	0	0	0
1	0	0	0	1
2	0	0	1	0
3	0	0	1	1
4	0	1	0	0
5	0	1	0	1
6	0	1	1	0
7	0	1	1	1
8	1	0	0	0
9	1	0	0	1

To make use of LEDs to display the count in BCD form, these are connected to the output terminals of the decade counter or to those of the latch (the 7475) as shown in Fig. 8.21. A 100 Ω resistor is connected in series with each LED to limit the current through it. The output to the next stage is taken from the output terminal Q_D. When the number changes from 9 to 0 the output Q_D goes low and the LED connected to Q_D shows that this is the case. The carry signal is thereby provided to trigger the next stage (go from units to tens through 9 → 10) even though the display is being held.

The outputs Q_D, Q_C, Q_B and Q_A may be reset to zero by momentarily disconnecting pins 2 and 3 of the 7490 from the zero volt line. For this purpose a reset push-button switch S_1 is provided (Fig. 8.21).

8.6.4 *Counting the pulses from a low-frequency multivibrator.* A suitable very low frequency (approximately 2 Hz) multivibrator is the 7400N module based on a quad 2-input NAND gate which provides an output between zero and +5 V. The circuit is the same as that shown in Fig. 8.15a except that the capacitance C_1 is 500 μF. The output from this multivibrator is connected to pins 3 and 4 of the 74121 (Fig. 8.22) where the Schmitt trigger input (pin 5) is disconnected (most easily achieved by unplugging the telephone dial) to avoid inhibiting the input to pins 3 and 4.

Appendix

The experiments of which details are given in this book are based largely on a set of 'modules' specially designed by Dr. K.J. Close. These 'modules' are now made and marketed by Irwin-Desman Ltd of Purley Way, Croydon, CR9 4QL. This so-called 'TEONIC' teaching electronics system (which includes instruction booklets) is based on 'modules' each of which contains on a plastic mount the essential active and passive components, the appropriate circuit diagram and terminals of a special design for very easy connection by electrical leads to power supplies and measuring instruments. Each such 'module' enables a range of experiments to be performed. Whereas the text of this present book is progressive independently of these 'modules', it is thought useful to give below the TS numbers and titles of the modules in the Teonic Sets because almost all the experiments described in this book can be undertaken readily by the use of these sets.

Set no. TS401 comprises:
401.1 Diode characteristics; 401.2 DC power supply unit; 401.3 Junction transistor in common-base connection; 401.4 Junction transistor in common-emitter connection; 401.5 Transistor tester; 401.6 Unijunction transistor (UJT); 401.7 Constant current source; 401.8 Emitter follower voltage stabilizer; 401.9 Free-running

multivibrator; 401.10 Introduction to silicon-controlled rectifiers (SCRs).

Set no. TS402 comprises:
402.1 Field effect transistor (FET); 402.2 Common-emitter amplifier and an equivalent circuit; 402.3 Introduction to operational amplifiers; 402.4 Voltage stabilizing circuits (using difference amplifier); 402.5 AC phase control with silicon controlled rectifiers; 402.6 Schmitt trigger circuit; 402.7 Voltage sweep circuits; 402.8 Multi-vibrators using field effect transistors; 402.9 Bistable multivibrator using transistors; 402.10 Bistable circuit using SCRs; 402.11 Programmable unijunction transistor (PUT).

Set no. TS4403 comprises:
403.1 Differential amplifier using field effect transistors; 403.2 Integrated circuit operational amplifiers; 403.3 Monostable multivibrator; 403.4 Voltage stabilized power supplies; 403.5 Waveform generators using NAND gates; 403.6 Voltage-to-frequency converters; 403.7 Wien bridge oscillator; 403.8 An optically-coupled isolator; 403.9 Decade counter and cold-cathode number display tube; 403.10 A monolithic waveform generator.

Index